Repeatability, Reliability, and Scalability through GitOps

Continuous delivery and deployment codified

Bryan Feuling

BIRMINGHAM—MUMBAI

Repeatability, Reliability, and Scalability through GitOps

Group Product Manager: Wilson Dsouza

Publishing Product Manager: Rahul Nair

Senior Editor: Arun Nadar

Content Development Editor: Sayali Pingale

Technical Editor: Sarvesh Jaywant

Copy Editor: Safis Editing

Project Coordinator: Shagun Saini

Proofreader: Safis Editing

Indexer: Rekha Nair

Production Designer: Alishon Mendonca

First published: April 2021

Production reference: 3140521

Published by Packt Publishing Ltd.

Livery Place

35 Livery Street

Birmingham

B3 2PB, UK.

ISBN 978-1-80107-779-8

www.packt.com

To my wife, Julia, who has been extremely supportive throughout this whole process. To my children: Sadie, Silas, Levi, and Evangeline, who have all been encouraging throughout this writing process. And most importantly, Soli Deo gloria!

– Bryan Feuling

Foreword

Having been a hero for IT organizations, I realize the battles that are waged to sustain the systems we take for granted. They are the enduring talent that knows the exact script to run, or commands to execute, to bring back our desired apps. That life is one that puts family and self care on the back burner due to the impact of constant weekend and evening escalations. Enlightenment and healing come from effective software engineering practices such as GitOps. Talent heroics is perpetuated by the belief that tasks themselves matter more than the value they create. GitOps supports the idea that an SME can create a process whereby the outcome is attained, consistently, without their direct involvement or loss of valuable time. Once you realize the scaling and sustainability that brings, you begin to focus on self-development and continuous improvement. GitOps is the practice of leaving behind worthy evidence of your passage at an organization, building process-based outcomes, using the skills that make you an SME, that persist beyond your move to bigger, more valuable challenges. Sustaining systems, creating services, based on very effective code-based processes. This book is targeted at those individuals in the hope that it will provide them a well-articulated path for enlightenment and healing from their talent heroism.

– *Shlomo Bielak* (CTO at Benchmark Corp)

Contributors

About the author

Bryan Feuling began his tech career as a help desk technician for a Fortune 200 company. His experience includes database administration, application development, automation engineering, and more. He has worked with hundreds of companies to help them avoid the same issues, pains, and engineer burnout that he has seen and still sees in the industry. Bryan has helped many companies and engineering teams implement GitOps practices at scale.

About the reviewer

Ravi Lachhman is a technology evangelist focusing on the cloud-native and CI/CD space. Ravi has held various engineering, evangelism, and sales roles at AppDynamics, Mesosphere [D2IQ], Red Hat, IBM, and Deloitte. Ravi is an organizer at DevOps Days Atlanta where he acts as the SRE talk track chair and an organizer for {unscripted} conferences. In Ravi's free time, he enjoys traveling and exploring new restaurants and cuisines. Ravi resides in Atlanta, Georgia with his Siberian Husky, Kofi, and is a graduate of Georgia Tech.

> *– Everyone learning more about CI/CD and helping further the craft of software delivery, this book is for you. Special shout-out to the author, Bryan, for furthering GitOps for everyone. Lastly, thank you to my Siberian Husky, Kofi, for waiting patiently for walks while I was reviewing this great book.*

Table of Contents

Section 2: GitOps Types, Benefits, and Drawbacks

4

The Original GitOps – Continuous Deployment in Kubernetes

5

Purist GitOps – Continuous Deployment Everywhere

6

Verified GitOps – Continuous Delivery Declaratively Defined

7

Best Practices for Delivery, Deployment, and GitOps

Section 3: Hands-On Practical GitOps

8

Practicing the Basics – Declarative Language File Building

9

Originalist Gitops in Practice – Continuous Deployment

10
Verified GitOps Setup – Continuous Delivery GitOps with Harness

11
Pitfall Examples – Experiencing Issues with GitOps

12
What's Next?

Other Books You May Enjoy

Index

Preface

GitOps has become a very popular term in the CI/CD and DevOps industry recently. From its inception a few years ago, GitOps has grown and mutated to the point where it has become difficult to understand and even difficult to implement. This book is intended to clear the air surrounding GitOps, especially considering the many different tools and vendors that claim to be GitOps tools without a clear definition of what they mean. Understanding the different GitOps types on the market today, what the benefits and drawbacks are of each of those types, and what anti-patterns to look out for, are all important when considering GitOps for your organization. By the end of this book, you will have the understanding required to adopt the appropriate GitOps style and direct your team to a more repeatable, reliable, and scalable deployment and delivery practice.

Who this book is for

This book is intended for those in engineering leadership that are looking to adopt a more repeatable, reliable, and scalable deployment and delivery practice. Anyone who has either already adopted GitOps or is researching GitOps for their team or organization will benefit from this book as well. The overarching goal of this book is not to promote a specific practice or tool, but rather to show the benefits and drawbacks of the different GitOps styles and how a team, group, or company can most benefit from GitOps practices.

What this book covers

Chapter 1, *The Fundamentals of Delivery and Deployment*, explains that to be able to understand GitOps, especially with a wide range of audience familiarity with DevOps and Declarative Languages, a refresher is required to set the proper foundation.

Chapter 2, *Exploring Common Industry Delivery and Deployment Practices*, after having set a foundational understanding of terms and general concepts, analyzes how others in the industry are solving for delivery and deployment, and discusses the desire to make things "continuous."

Chapter 3, The "What" and "Why" of GitOps, with all of the fundamentals, analyzing, and general concepts covered and understood, dives into what GitOps is and why it matters to the engineering team and business, especially in the desire to have "continuous" practices.

Chapter 4, The Original GitOps – Continuous Deployment in Kubernetes, explains that when the concept of GitOps was first introduced, the intended target was for Kubernetes. This chapter covers why and contains a high-level view of what it looks like in practice.

Chapter 5, The Purist GitOps – Continuous Deployment Everywhere, shows that purest GitOps is the idea that GitOps should work for any deployment type using declarative language, regardless of the technology stack used.

Chapter 6, Verified GitOps – Continuous Delivery Declaratively Defined, explains that verified GitOps is the only concept that looks to define Continuous Delivery as code.

Chapter 7, Best Practices for Delivery, Deployment, and GitOps, contains best practices for the business or team with regard to CI/CD, allowing tool interchangeability.

Chapter 8, Practicing the Basics – Declarative Language File Building, discusses building declarative language files, which is the most basic practical requirement for GitOps.

Chapter 9, Original GitOps in Practice – Continuous Deployment, builds on the experience gained from the last chapter to explain how to set up VS Code and Minikube, and use file-save triggers.

Chapter 10, Verified GitOps Setup – Continuous Delivery GitOps with Harness, explains how to declaratively define the Continuous Delivery process, which provides process repeatability and scalability.

Chapter 11, Pitfall Examples – Experiencing Issues with GitOps, discusses some of the common pitfalls when trying to use GitOps for the first time.

Chapter 12, What's Next?, summarizes the fundamentals, philosophical ideals, best practices, and practical examples.

To get the most out of this book

Although there are sections of this book that walk through setting up different tools and examples, the majority of this book is intended to address the methodologies and purpose of GitOps. The intended audience is anyone in engineering leadership that wants to use industry practices to enforce a repeatable, reliable, and scalable deployment and delivery process. The minimum knowledge requirement for this book is some familiarity with a software development life cycle.

Software/hardware covered in the book	OS requirements
macOS Big Sur	Helm 3.5.0
VSCode 1.54.3	Argo CD 1.8.5
Minikube 1.17.0	VirtualBox 6.1.18
Homebrew 3.0.5	

Download the color images

We also provide a PDF file that has color images of the screenshots/diagrams used in this book. You can download it here:

```
https://www.packtpub.com/sites/default/files/
downloads/9781801077798_ColorImages.pdf
```

Conventions used

There are a number of text conventions used throughout this book.

DevOps Analogy (Italics): Indicates the analogy of a DevOps team going through the process of adopting Continuous Deployment and Continuous Delivery. This analogy can be found at the beginning of every heading in each chapter, followed by an explanation of the analogy.

`Code in text`: Indicates code words in text, database table names, folder names, filenames, file extensions, pathnames, dummy URLs, user input, and Twitter handles. Here is an example: "The automation of the tool runs `helm install` or `kubectl apply`."

A block of code is set as follows:

```
# BASIC STRUCTURE
name-1: "hello-world-1"
description-1: "say hello to the world 1"
count-1: 1
```

Bold: Indicates a new term, an important word, or words that you see onscreen. For example, words in menus or dialog boxes appear in the text like this. Here is an example: "Select the **Disable** button."

> Tips or important notes
> Appear like this.

Get in touch

Feedback from our readers is always welcome.

General feedback: If you have questions about any aspect of this book, mention the book title in the subject of your message and email us at customercare@packtpub.com.

Errata: Although we have taken every care to ensure the accuracy of our content, mistakes do happen. If you have found a mistake in this book, we would be grateful if you would report this to us. Please visit www.packtpub.com/support/errata, selecting your book, clicking on the Errata Submission Form link, and entering the details.

Piracy: If you come across any illegal copies of our works in any form on the Internet, we would be grateful if you would provide us with the location address or website name. Please contact us at copyright@packt.com with a link to the material.

If you are interested in becoming an author: If there is a topic that you have expertise in and you are interested in either writing or contributing to a book, please visit authors.packtpub.com.

Reviews

Please leave a review. Once you have read and used this book, why not leave a review on the site that you purchased it from? Potential readers can then see and use your unbiased opinion to make purchase decisions, we at Packt can understand what you think about our products, and our authors can see your feedback on their book. Thank you!

For more information about Packt, please visit packt.com.

Section 1: Fundamentals of GitOps

This section is designed to provide a foundation that you can build your best practices on in regards to continuous deployment, continuous delivery, and GitOps.

This section comprises the following chapters:

1
The Fundamentals of Delivery and Deployment

Any company that builds and maintains applications is automatically concerned with repeatability, reliability, and scalability. In fact, some of the main metrics that are monitored on an application are directly related to these operational concerns. Understanding the basics and history of the industry when attempting to accomplish the ultimate trifecta of software administration is paramount to learning from the issues of the past.

In this chapter, and throughout this book, you will embark on a journey with a DevOps team as they attempt to conquer the deployment and delivery world. By experiencing the pains, bottlenecks, and setbacks with the DevOps team, you will understand how the industry has evolved, and what needs to be accomplished in order to succeed.

In this chapter, we're going to cover the following main topics:

- How did we get here?
- What is a deployment process?
- What is a delivery process?
- What makes any practice continuous?

How did we get here?

It's 8 a.m. on a Saturday and the release party's post-mortem has finally been completed. Throughout the release, every encountered issue resulted in a Root Cause Analysis process. Once each of the RCAs were done, the release team would then create and assign tickets as needed, resulting in action items for the different teams in the Engineering organization. With the post-mortem being completed, the release team can hand-off the monitoring of the production application to the weekend support team and head home.

The final production servers were upgraded with the new application release at around 3 a.m. that morning with all of the application health checks successfully passing by 3:30 a.m. And yet, even with the early morning finishing time, this is a significant improvement when comparing it to the release parties of a few years ago. Previously, the applications were released every 6 to 12 months, rather than the quarterly release cadence that the company is currently on.

Their company had hired a consulting agency to advise them on how to improve their application's mean-time-to-market and reduce their production outages in order to meet business initiatives and demands. The outcome suggested by this consulting agency was to release the application more frequently than once or twice a year. As a result, the releases have been quicker and less prone to error, which the business has taken notice of. The release parties still require pulling an all-nighter, but the previous release parties were more like all-weekenders or longer.

The on-call engineering team still has to be brought in for every release, but at least they aren't required to be a part of the release party for the entire time. And the most recent release only required a conference bridge for about 4 hours to solve issues with the underlying code or provide quick fixes. Overall, the operations team, infrastructure team, network team, and security team were able to solve most of the issues that showed up, which accounted for significantly more confidence in the newer release cadence.

The different teams should be able to accomplish the backlog of issues before the next release. But the team with the largest issue backlog were the systems administrators , who build, integrate, administrate, and troubleshoot the many different tools used during the releases.

After 12 straight hours with over 15 members across a host of different teams, the release party was complete. When considering the time associated with the attempt to improve the process throughout the quarter, as well as the actual release itself, it is not difficult to run the mental math on the associated costs. The teams need to figure out a way to make the releases more reliable, repeatable, and scalable.

This analogy is all too familiar for many who have been involved in the engineering side of a business during the Waterfall software development life cycle days. When applications were first made available as a **SaaS** (**Software as a Service**) solution,the common release cadence was an annual release. Throughout the year a company would deploy small release, often called patches, which mainly consisted of hardware, software, or security updates.

Since the yearly update was essentially releasing a brand-new product, the release process required significant involvement from every team across the entire engineering organization. The release was a major event , often taking an entire weekend, or longer, from every team available. Many in the industry had dubbed this event a release party. Each release party included significant amounts of caffeine and food, which accompanied a host of people hunched over their laptops as they watched the output of the release on a massive projector screen. Yet the worst part of this whole scenario was that this was the expected release style for every company at the time.

The quarterly release cadence was a novel idea that revolutionized how companies would develop and test their code. The code changes were smaller in nature and the teams evolved their thinking from *a new product every year* to *a new subversion every quarter*. Some user experience changes may be introduced, but most of the user experience in the application would remain the same from release to release. Another major benefit to the increased release frequency was the significant reduction in lead time, which is the time it takes to go from a feature being requested to being available in production.

Alongside the release parties were two very important processes when issues would arise during the release:

- Root-cause analysis (RCA)
- Post-mortem

An RCA would occur anytime there was a significant issue in production that would halt or severely affect the functionality or availability of the application. Often, the RCA process would start with the teams analyzing what was wrong, fixing the issue, validating that the fix worked, and then documenting how the issue arose and what the root cause was. Every release party would result in at least one RCA taking place, and would exponentially increase in number relative to the total amount of production servers involved in the release party.

The post-mortem was a retrospective process after the release was completed and the teams were confident in production operating as expected. The release captain would gather any and all information related to RCAs, bugs, errors, and so on, and create the required documentation and tickets. At the end of the post-mortem, a weekend support team would be briefed on the release party outcome and any items needing to be monitored.

The desire to automate the release of the application had been a central focus of every engineering organizations for years Automation was seen as the best way to enforce reliability and repeatability into the release process, and most of the common tools in use today were created with the intention and purpose of release automation. These tools, and really the underlying processes they address, intend to solve two major concepts in the software development life cycle: **deliveries** and **deployments**.

What is a deployment process?

10 p.m. on Friday was when everything started falling apart. The Q2 release party started a few hours ago with the entire operations team and a few members of the infrastructure and network team in attendance. Routing customer traffic away from the initial test server to allow for the upgrade went as expected. This process was recently automated through some network management scripts that the systems administration and network team worked on. The idea was that all new traffic should be routed away from the initial server while allowing the customer sessions that were currently using the server to continue until they disconnected. After all the user sessions were completed, the server was removed from the load balancer and the release process could start.

The infrastructure team had a bootstrap script already built out to automatically configure the server. Sometimes this process involved tearing down the whole server and rebuilding it, while other times the release required some simple software updates to be completed before the hardware was ready for the application release. The new release wouldn't require an entire rebuild of the server this time. However, since the last release was 3 months ago, they did have to patch the server, add a new application stack version, and make sure that other configuration requirements were set accordingly. The entirety of the infrastructure process took about an hour for the first server, which would then be repeated for the other servers so that the bootstrapping time would be reduced for the rest of the fleet. As more customers were acquired, the total number of servers in production had grown. To avoid downtime for the production environment these servers were grouped together into pools, which could then be individually targeted for stopping, upgrading, and restarting as needed.

After the initial server was bootstrapped by the infrastructure team and validated through some basic quality and security tests on the server, the operations team would then start the application release process. It was just after 7 p.m. when the operations team started the release process, also known as a deployment, by copying the ZIP file from the production network share to the server. The file was then expanded, into a mess of files and folders which contained system services, application files, and a rather daunting INSTALL_README.txt *file. This README file detailed all of the required install steps and validation checks that the engineering team documented for the operations team to execute.*

With the install instruction file open on one screen and the terminal open on another, the install process could start. That is when everything went wrong.

Although the deployment testing in the staging environment had some issues because of missing requirements, those were documented and added to the install process. But what the operations team didn't know was that the server bootstrapping script had reset all of the network configuration files and all of the aplication traffic heading out of the server was being redirected back to itself. As the deployment went through, the application ZIP file was able to get pushed to the server, the filesystem was set up as needed, and the required system services began running. The script used to test the health of the application showed all successful log messages. However, when the script to test the interaction between the application and the database was run, the terminal output showed only connection errors. It took the team over an hour to get everything copied over, stood up, and tested before the network errors were discovered. The release party had come to a grinding halt.

The operations team was in full-on panic mode and the first RCA process had started. If they could not figure out why the server was not able to talk to external machines within the next hour, they would need to tear down the whole server and start over again. While one person from the operations team collaborated with the network and infrastructure team, another operations team member would retrace every action taken since the infrastructure team had finished their tasks. After 30 minutes of the network team analyzing all traffic related to the new server and the desired databases, they could not find any reason as to why the server could not reach the database. The Infrastructure team was checking to see if the server had been properly added to the domain and that no other machines were using the same hostname or IP address. The operations team had engaged the on-call engineering team and started a troubleshooting conference bridge for the data center support team to join.

It wasn't until a few minutes after midnight that the network team found the networking loopback issue on the server. The outcome of the RCA process found that the server bootstrapping script was the culprit, which was then altered to avoid the issue in the future. The server was now passing all health checks and the operations team could move on to the next server in the pool. Within an hour, the rest of the server pool had been fully upgraded without an issue. Almost two hours later, all server pools were upgraded and reporting healthy. The post-mortem process could begin now that the new application version was out in production and operating as expected.

A release party would always start with an initial test release into production, known as a deployment. At a high level, a deployment process solely concerned with is copying an artifact from a designated location to some endpoint or host. In the case of the quarterly release party, a deployment consisted of pushing or pulling the artifact to a designated *test* server in the production fleet. This was a common method used to avoid production downtime by preventing unknown production-specific nuances from negatively affecting an environment-wide deployment. The log output and application metrics for the test deployment were heavily scrutinized in an attempt to catch any hint of an issue.

The deployment on the test production server would typically require some bootstrapping process to enforce a common starting point for all future deployments. If all deployments started with the same server configuration starting point, then theoretically, nothing should be different as the deployments moved from one server to the next.

Once the deployment was complete, before traffic would be allowed on the application, a set of tests would be executed.These health checks would validate many different requirements, such as external service connectivity and buisness-critical functionality. After that initial production server was completed and the validation tests had passed, the rest of the initial server pool would be put through the same process. After the initial server pool was upgraded, it would be added to a load balancer with a set of load and smoke tests validating that the application, servers, and networking were operational.

Finally, after that initial server pool was completed, production traffic would then be routed appropriately. The next server pool in the queue would then follow the same process to test for deployment consistency. Once the release team had confidence in the process, they could upgrade multiple server pools in.

Even with a release party as intense as the one described happening on a quarterly basis, it was not the only release party for the engineering teams. Two other major deployment events would take place throughout the quarter: one for deployments of patches and hotfixes across the different architecture layers, , and the other for all non-production environments.

The first deployment that would happen throughout the quarter was when the engineering teams would need to test their code. The testing of the code started by packaging the code into an artifact and uploading it to a network share. Once the artifact was on the network share, the engineering team could then deploy their artifact to a designated server in the development environment. Usually, the engineer would have to run a bootstrapping script to reset the server to a desired state.

The process of deploying an artifact to a development environment had to be relatively repeatable , since the deployment frequency was either daily or weekly.. When the engineers believed that they had an artifact ready to be released, they would then deploy it to a test server in the **Quality Assurance (QA)** environment for further evaluation.

Along with all of the QA processes and testsneeding to be run, the team would need to start building out the `INSTALL_README.txt` file for the production release. The QA team would send testing feedback to the developers for any required fixes or improvements. After a few rounds of feedback between the teams, the most recent artifact version would be promoted to a release candidate. The teams would then focus on the deployment process for the next release party. The handoff of the artifact from the developers to the operations team would happen about a month before the release party. Often described as "throwing it over the wall", the developers would have little to no interaction with the artifact once it was passed on the operations. The operations team would then spend the next month practicing the deployment for the release party.

The other major deployment events taking place throughout the quarter were the patching and hotfix releases.

Similar to the development deployment process, the patching deployment process would be executed against lower-level environments for both testing and repeatability. The major difference was that these deployments would take place outside of typical maintenance windows.

The initial set of releases would start with the development environment, allowing for significant testing to take place. This would prevent regressions from affecting higher-level environments, such as QA or production. Once the deployments to the lower-level environments were repeatable, the teams would designate an evening or weekendfor the deployment to take place. Similar to the release party, one server would be removed, patched, restarted checked and then made available for users. Assuming everything behaved as expected with the patched application, the rest of the servers would be put through the same set of tasks.

A deployment is an essential process focused on getting an artifact into an environment. The more a deployment process can be automated, the more repeatable, reliable, and scalable the deployment process becomes.

What is a delivery process?

The time between the release parties seemed to get shorter with each quarter, especially when the requirements continued to grow. The different engineering groups were becoming smaller as engineers were burning out and leaving. The system administrator team needed to quickly figure out what had to be fixed and what could be automated. But after every release party, the backlog of work items continued to pile up, resulting in a mountain of technical debt.

The recent move from semi-annual releases to quarterly releases required a massive operations team overhaul. Some of the more senior infrastructure and engineering team members moved to a new group focused on automating the release process. But before they could automate anything, they needed to understand what a release actually consisted of. Since no team or person knew all of the different release requirements, the next quarter was spent on research and documentation.

What they found was that every step related to the lifecycle of the artifact, from development to production release, needed to be automated. This lifecycle, which included all pre-deployment, deployment, and post-deployment requirements, was considered a delivery. What they were surprised by was finding out that one execution of the delivery process took 3 months to complete. To make matters worse, every step in the delivery process had never been documented or defined. Every member of every team that was involved in the delivery process had a different reason for why each of their steps were required.

After the team had completed their initial documentation process and also after experiencing the recent release party, they were ready to start automating. But what they were unsure of was whether or not they should build out the process or if they should look for a tool to do it for them.

At first, they researched the available tools in the market that might give them a foundation to build the new process on top of. One issue they found was that the more common tools were related specifically to either Windows or Linux, but not both. The other tools that they had found were scalable across the different systems, but they required significant ownership and hosting requirements. Considering the short timeline and technology requirements, any tool that supported multiple systems and could be highly customized and extended through scripting would be best.

The system administrator team decided that it would be best to split up and tackle different requirements. Some of the team focused their attention on running tooling proof of concepts. The rest of the team would focus on building scripts to support the rest of the engineering organization. The initial iteration of the new deployment process would be focused mainly on the ability to build and execute the automation scripts. Once that was built out, the next iteration improvement would focus on turning the scripts into templates for scalability.

The first piece to automate was the deployment of the release candidate artifact to the test server. This required bootstrapping the server (resetting it to an optimal state, adding the desired environment variables, adding any required software or patches, and so on). Then they would pull down the artifact from a designated network share, expand it, and upgrade the application on the server. After that was completed, the automation process could then email the QA team for them to start their testing requirements.

With the development deployment completed, automation process would be directed at a staging environment. This environment contained multiple production-like servers , allowing the operations team to practice the deployment process. Automating this requirement, meant that the deployment scripts had to affect the network configuration, as well as the application server. After the server was bootstrapped and reset, the testing process could then be run to validate the application health. But, to make the automation more scalable, the team would need to have the scripts run remotely, pushing down that artifact instead. The remote execution behavior would allow for a larger deployment set to be run in parallel.

The last part that the team wanted to automate was the post-deployment verification step. This step could be run remotely and would pass data to the application. This would allow for both a network connectivity check and a desired functionality check.

The team would need to test out the automation process in production

One of the biggest issues that any engineering organization must deal with is technical debt. Technical debt is the cost of any rework that is caused by pursuing easy or limited solutions. And what makes technical debt grow is when engineering organizations work as a set of disparate units. This causes *compound interest*, since no central team will be able to maintain the cross-team technical debt.

ventually, the creditor comes to collect the technical debt and, depending on the communication styles of the teams and how long the debt has been avoided, the hole is almost too deep to climb out of. Technical will often go unnoticed for a few years few years until the creditor comes to collect. With regard to a engineering organization's technical debt, the creditor is often business or market requirements. Most recently, smaller and more agile start-ups disrupt the market, taking market share, and causing panic for the bigger players.

With the potential of technical debt resulting in a form of technical bankruptcy, many companies make radical decisions. Sometimes they decide to create new teams focused on a new company direction. Other times they will replace the management team for a "fresh perspective". But in any case, the goal is to repay the technical debt as quickly as possible.

A common place to find technical debt is in engineering supporting processes. For the system administration team in the analogy, most of the technical debt was associated to their release practice. Although they had a relatively automated deployment, they found that most of the manual steps were things that occurred before and after. As as result, the team realized that the biggest source of technical debt was their delivery process.

Any desire to automate a process must first start with a requirements gathering process. With the requirements gathered, a team can then pursue a minimally viable product. Part of the requirements gathering process is being able to define what the immediate needs were and which capabilities could be added at a later time. A minimally viable product is exactly as it sounds, a product that meets minimal requirements to be viable. In the analogy, the items that would be required in the MVP were server bootstrapping, artifact deployment, and network management. These functionalities would have the highest level of impact on the technical debt and also on the main problem areas throughout the current delivery and deployment process. Features such as running and evaluating tests, approval steps, dynamic environment creation, and traffic manipulation would be brought in over time.

Building, testing, and iterating are the common development cycles that any engineering team will need to go through. But the moment that any process is automated, the team responsible for the automation will need to consider scalability. Once another another automation use case is discovered, the automation tooling must be scaled to accomodate. The term often associated with scaling across use cases is known as onboarding. And a requirement to onboard other use cases or teams immediately develops a need for a central management team. That team will have the goal of supporting, improving, troubleshooting, and onboarding the solution for the foreseeable future. Eventually, the automation process becomes a core support tool that must be reliable, scalable, and offer repeatable outcomes.

What makes any practice continuous?

It's been about 2 years since the quarterly releases were mandated, and one year since the systems administration team had first been formed. The most recent release party resulted in a very low impact release. The engineering organization was able to significantly reduce the years of ignored technical debt. The company's product roadmap, promising a new product every year, recently went into effect. As a result, the automated delivery pipeline was now supporting two separate products in a repeatable, reliable, and scalable way.

A design choice that the systems administration team had pursued was building out an imperative configuration method for the automation solution. Looking back, the team realized that they would have had an easier time scaling through declarative methods instead. The need to quickly move away from the previous high-touch process drove the team to make some rushed decisions. Although variables could be added to the execution process, every new product required the team to clone and change the automation scripts to be successful. The new automation solution could only scale through this cloning process, resulting in heavy administration and storage costs.

The system administration team realized that for a tool to support their scaling requirements , the tool would need to support declarative configurations and executions. This requirement became abundantely clear as the architecture support requirements grew and changed. For now, the team could convert their scripts into more declarative templates to implement a short-term scalable solution. The system administrator team needed to get a new process in place, and fast. The previous process was error prone and highly manual, resulting in a mountain of technical debt. But to develop the MVP in time, the team took a risk of piling up technical debt of their own. They assumed that they would have enough time to fix and refactor the solution to pay off the technical debt. But this time, the creditor came back much sooner than expected.

Business executives were impressed with the lack of issues relating to the recent product launches. The reliability of their main product was better than ever. In fact, these recent changes resulted in the company gaining significant market share. The success of the two products in the market meant that the company could double their efforts. And to ensure that the recent success would continue, the company executives decided to hire a new Chief Digital Officer. This new CDO was well known in the industry for implementing signficant change in engineering. One of the major changes that the CDO was bringing to the company was adopting DevOps practices.

The following month saw a host of changes across all of engineering. Every team was now required to attend DevOps training and enablement. Anyone in engineering leadership was required to read a lengthy handbook on DevOps. The different teams were now also being tasked with documenting their current process, as well as anything that they were working on. Each team would also be incentivized to learn about Git, containerization, and different continuous practices.

The infrastructure, network, and security teams were tasked with learning about containers, container orchestration, and cloud infrastructure. The operations team became the DevOps team, and the system administration team became site reliability engineers.

These changes required the teams to migrate from current process to DevOps practices. The development team had more time granted for their migration requirements. But the DevOps and SRE teams were required to rapidly migrate current platforms over to cloud native ones.

The significant shift in direction towards DevOps and cloud native technology resulted in a major staff change. Some of the more senior engineers left the company, while a host of new hires brought in fresh perspectives. The goal was to get the company out of the Waterfall software development life cycle method and into the continuous world of DevOps. The CDO wanted an integration, deployment, and delivery process that was executed at least once a day.

Many companies that existed at the time when the Waterfall software development life cycle was the industry best practice have been confronted with this need to change. The shift from Waterfall methods to Agile, and now to DevOps has rocked the engineering industry. The ability to execute a delivery or deployment process at any time seemed too risky. The perspective was that only the largest companies with the most money and the largest engineering staff could achieve these extreme capabilties.

The DevOps and SRE teamsrealized that the best way to support the new DevOps requirements was to rebuild their automation solution. They would need to set up a best practice process for the developers to use. This included the tools, solutions, platforms, and steps needed to enable continuous delivery and deployments.

Different members of the DevOps and SRE teams would still need to maintain the old process in a weekly rotation schedule. Others on the team would work on setting up the new platform in the desired cloud provider and learning the steps to get an artifact there. After choosing a container orchestration platform, the DevOps and SRE teams needed to work on automation. Configuring and scaling the new platform required the teams to learn how to use Infrastructure as Code. A declarative method of delivery and deployment is one of the most scalable options for a DevOps practice. Most major platforms today natively support declarative configuration practices. This support allows for teams to easily adopt the platform and scale it out to meet their needs.

After the platform was set up and an artifact was deployed to it, the teams started looking at different templating options to make the administration requirements light. This led to a bake-off across different declarative deployment styles, some natively built into the platform and others leveraging an underlying templating engine. As the teams got closer to the desired end state, they had to find a new tool that would enable and assist them in the cloud-native world. The two main questions they needed to answer now were the following:

- Do they need to support the old applications as well as the new?

- Should they look for a continuous deployment tool or a continuous delivery tool?

Summary

This chapter presented a foundational understanding of what a delivery requires, what a deployment is, and how the industry arrived at the state it currently operates in.

A common misunderstanding in the industry is that if the company is using a continuous integration and/or continuous delivery tool, or if their application is releasing at least once a week, they are operating in a continuous manner. This chapter laid out that a continuous process is defined as one that executes at least once a day, preferably more.

In the next chapter, the DevOps team will explore common industry trends and tools related to software delivery and deployment, as well as how to test their automation process.

2
Exploring Common Industry Delivery and Deployment Practices

The analogy in the last chapter followed a company through the beginning of its digital transformation from error-prone quarterly release parties to DevOps and continuous SDLC practices, with the newly hired CDO driving the initiative.

This chapter will continue the analogy of following the company through the transformation process as they research tools, processes, and practices that are common in the industry and how they will implement them.

In this chapter, we're going to cover the following main topics:

- Common industry practices for deployment
- Common industry practices for delivery
- Common tools used for deployment and delivery
- Needs versus wants
- The automation test

Common industry practices for deployment

After almost 2 months of DevOps enablement, the DevOps and SRE teams were ready to implement the different continuous practices. However, migrating to the new process would not be an easy task for their developers. Getting the users to adopt a new practice as well as a new platform would be difficult. The best way to enforce standards and prevent requirements overload would be to build a highly configurable solution. The goal would be to have the developers only provide artifact and environment configurations. The rest of the requirements would be enforced by the solution itself.

One of the recent industry trends has been to leverage declarative language files to provide configuration requirements. A declarative language, such as YAML, JSON, or XML, leverage a key:value *style, often in a nesting layout for easy data storage, access, and readability. During their research, the teams discovered that most tools that were cloud native would use either JSON or YAML as their declarative language of choice. The team also found that, in many instances of using declarative files, users would also leverage templating to scale the configurability of the files. Some users would leverage jsonnet for JSON templating and go-templates for YAML templating. However, they also found templating capabilities with different open source projects, such as Jinja, and Mako, but those seemed less popular.*

Since the developers would need to maintain their own configuration files, a set of standards was required. The DevOps and SRE teams needed to select their desired declarative language. But, that decision would be impacted more by what templating style the DevOps team preferred. The cloud infrastructure team had already decided on which cloud provider to leverage. They also selected the cloud provider's implementation of a popular container orchestrator. The use of these solutions would allow the cloud infrastructure team to offer an entire platform service to all engineers. The DevOps team could then focus less on the infrastructure and more on the application requirements.

With the cloud platform and container orchestrator selected, the teams needed to define the developer process. The solution that the DevOps team would administer would allow the developers to host the configuration file in their Git repositories. The solution would then then use the file for the deployment and delivery.

One major piece of functionality that the DevOps team would need to configure is cross-regional deployments. The companies customer base was located across the globe, meaning that the application had to be available in their regions. To ensure that the application deployment was succssful, the team would deploy to a User Acceptance Testing environment first. This environment would be located where the majority of customers were. Different customers would be able to test out the new features before the features would replicate to other regions. This release style would result in two main deployment phases. Because the core set of resource files was templatized, the developers would not need a unique deployment method. Any lower environment specific configurations would be provided through the configuration files. This allowed the developers to test production-like processes and features across non-production environments.

The world of DevOps processes and tools has exploded in recent years, with every possible company, vendor, and group trying to capitalize on the massive market opportunity. Some tools market themseleves a a full platform, or by being open source, and others will leverage artificial intelligence and machine learning as a differentiation. Most companies that have been around for a few years are either just starting or in the middle of their DevOps journey.

Any company that develops its own application has a defined deployment process in place. But the ability to execute that process in a continuous manner is often very difficult. Some of the automation platforms that have been in use for a while were defined imperatively with many manual requirements. These inherent limitations make increasing deployment frequency difficult. Every engineering orgnaization has become focused on achieving continuous practices, such as integration and deployments. But the major hurdle to achieving continuous practices is not actually related to the tools used. Rather, the hurdle is trying to guarantee repeatability and reliability as the frequency increases. But even if a team can achieve a repeatable and reliable continuous process, making it scalable adds an extra layer of complexity.

Most teams that have successfully implemented a scalable, repeatable, and reliable process have had to address common hurdles. Since every application has unique configuration requirements, the developers must have an easy way to declare them. Not only will every application have unique requirements, every customer and region will have unique requirements as well. To achieve a truly scalable, reliable, and repeatable process, an execution engine must allow for all of these requirements to be applied correctly.

Two general methodologies exist for solving user interaction with the execution engine:

1. Since DevOps practices often focus on operation processes *shifting left* to the developers, those developers should be free to use whatever tools they want as long as they are productive. This results in the developers not only choosing which tools to use, but also all configuration, administration and troubleshooting of the tools.

2. Developers should focus solely on developing their code. A DevOps team should build and maintain a happy path for developers to use. All business requirements are enforced automatically without adding to the developer workload. This means that the developers will only be required to maintain a simple configuration file with limited variables, making for a simple, easy, and opinionated process for all.

When a DevOps team needs to decide which path would be the easiest to achieve the desired end goal, they must also consider any company-mandated timeframes. Because of these requirements, DevOps teams are often confronted with whether they should build or buy a platform. In either case, it is important to start by understanding what functionality and support is required. Developers with too many requirements will have lower production rates. However, developers with too little configurability will have less innovation. The ideal platform will increase developer productivity and innovation.

With companies becoming increasingly more software focused, developer productivity and innovation drive success. Allowing a developer the configurability and freedom that they need to improve the rate of innovation is essential.

The most common practice in the DevOps industry, regardless of the user interaction methodology chosen, is leveraging configuration files. Every common tool available, whether open source or not, has configuration file support. And when those configuration files can be stored and maintained in a Git repository, the scalability of the tool becomes significantly easier.

Common industry practices for delivery

The deployment process had been decided, along with the desired end user interaction method. The DevOps and SRE teams would have the developers store their configuration files in a Git repository. This would allow for easier scaling as teams would be onboarded. But now the DevOps team had to gather other requirements that the execution engine needed to support or enforce.

Interviewing engineering leadership would be the best place to start the requirements gathering process. The leadership team knew which compliance and auditing processes were important. The company had to maintain an ISO27001 and SOC2 certification for them to retain many customers. Also, since the company had a financial application that some government entities would use, they had to undergo PCI and FedRAMP auditing.

Other requirements included support for highly-available systems, disaster recovery requirements, and a 99.99% uptime **Service Level Agreement (SLA)***. Breaking these SLAs, which unfortunately occurred about half of the year, resulted in losing customers and losing revenue. With a high importance being placed on the SLAs, the SRE team split from the DevOps team to focus solely on that. The SRE team would also focus on making the production environment highly-available with a disaster recovery pair. The DevOps team would focus solely on the deployment and delivery pipeline. Additionally, the SRE team would have to define and monitor the* **Service Level Objects** *(SLOs) and* **Service Level Indicators** *(SLIs) associated with the SLAs.*

The DevOps team also needed to figure out how other teams were defining and building out their areas of responsibilities. The cloud team had requirements that also needed to be automated, which they were doing through Infrastructure as Code tools. The security also had requirements to harden the cloud infrastructure as the platform was being built out.

The teams then interviewed the quality team to understand what type of testing would be required as new features or functions were developed. Recently, the quality team was focusing more on building out testing automation for the continuous integration practice. The quality team was also wanting to build out smoke tests and load tests to run in production. These tests would help with monitoring the SLIs for new deployments. The security team had different vulnerability scanning requirements that needed to be implemented. The desire was to leverage artifact scanning at the beginning of the delivery and application vulnerability scanning at the end.

At the end of the requirement gathering process, the DevOps team interviewed the developers. Although most of the developer's work would be segregated to the integration process, there were some tests that needed to be run. The developers also needed quick feedback in regard to the application health and any required fixes. Monitoring was always a blindspot for the development team and the new process needed to fix that. Although the compliance requirements would not allow the developers to have access to the higher-level non-production and production environment, they would still need a way to get a feedback loop from those environments.

To build out an execution engine, the DevOps team needed to understand which requirements were actual needs. The rest of the requirements would be added at some point, but the goal was to get an MVP platform. The next major part would be to start building out the execution engine to take in declarative files for the delivery and deployment instructions. The initial iteration of the delivery pipeline would be deployment focused. Multiple environments and multiple regions was a fairly large task to automate. The DevOps team also wanted to include automating the ticketing and approval process, if possible. Any other tasks would have to be manual for a time.

One of the most difficult and time-consuming portions of the SDLC process that every DevOps team has had to solve is the **Continuous Delivery (CD)** part of the CI/CD acronym. Every company that has their own application will have a deployment process defined already. But automating everything before and after the deployment is where the difficulty lies.

One of the biggest problems associated with the delivery process is that multiple teams are involved across the business, regardless of their involvement in engineering. The business-minded teams are concerned mostly with the outcome of their customer's product usage. Generating revenue, customer satisfaction, and brand recognition are their main concerns. And, as the DevOps movement's impact on business performance propagates through the different industries, companies will use DevOps-based metrics to evaluate their engineer's performance. The most common metrics being monitored are Developer Lead Time, Change-Failure Rate, Deployment Frequency, and Mean-Time-To-Restore. However, without a well-defined delivery pipeline, the metrics-gathering process is almost impossible.

Those on the engineering side of the company translate business requirements into metrics that directly correlate to their SLAs. All SaaS companies advertise that their application has a certain number of *9s* of up-time. What this means for their customer's is that unscheduled outages happen infrequently and for short periods of time. The available uptime of a product is the most common contractual metric for a business. Additionally, all companies with a SaaS application will advertise their uptime as a representation of quality and reliability. Breaking an uptime SLA can present a major challenge to a company. Customer's will make a direct correlation between a company's uptime and their reliability. As a result, broken SLAs will often lead to customer churn.

SLAs are not the main metrics that the engineering team is concerned about, but rather the SLIs and SLOs. An SRE teams main focus is on tracking SLOs. By monitoring the SLOs, the SRE team will be alerted when the SLIs operate out of a desired bound. To improve the reliability of a system, the SRE team will spend significant time keeping the SLIs within a desired range.

To achieve any business requirement, the DevOps and SRE teams must enforce three essential requirements for every delivery pipeline: Repeatability, Reliability, and Scalability.

Platform reliability is often related to the cloud infrastructure. Most infrastructure teams will leverage an Infrastructure as Code tool to enforce consistency. Once requirements are converted into code, the process immediately becomes repeatable. Code should always execute in an expected manner, known as idempotency. If an execution engine allows for configuration to alter behavior, then the solution is both repeatable and scalable. When leveraging a highly configurable execution engine against a highly-available infrastructure, then the platform becomes reliable.

The next major portion to consider for a delivery pipeline is enforcing quality and security. Quality and security need to be applied to the application delivery and user interaction. This requires collaboration with the security and quality teams to understand both the current process and desired process. User security within a delivery pipeline will focus mostly on admission control and auditability. Application security is most focused on vulnerabilities across the artifact, environment, and user interface. The quality team is usually focused on testing business requirements and new functionality. Any business requirements will typically have automated testing assigned to it. While new functionality testing will either be the result of developer-built tests or well defined desired outcomes. The quality team can become a bottleneck if the tests are not automated or if the tests do not provide full code coverage. Quality teams can build out tests that can be executed automatically, but a massive time saver is being able to evaluate the outcomes automatically. But even if tests can be executed and evaluated automatically, if the tests do not provide complete coverage, then manual steps will still be required. Building out full testing coverage that can be executed and evaluated automatically will result in a more reliable application.

Every team will have dependencies related to other teams in the organization. But the core group that all teams depend on are the developers. As such, understanding what requirements a developer has will directly impact the requirements of other teams. When building out a delivery and deployment pipeline, a developer will need a way to pass in configuration information. This is often accomplished through configuration files, which the developers will host and maintain.

Every team has requirements, and every team has a desired process. But the most difficult task when building out a delivery pipeline is discerning the difference between a true requirement and a desired state. The industry standard delivery process, regardless of team requirements, is automated deployments and manual everything else.

Common tools used for deployment and delivery

As the final documentation is compiled for the delivery and deployment requirements, the DevOps and SRE teams need to start building out pipelines. There is a desire to make a pipeline that can take in declarative files that only define the desired outcome. However, the team might have to settle for tools that are imperatively designed. This means that the users will need to explicitly define the **what** *and* **how** *of an execution process until the solution evolves over time. The main issue with an imperative solution is that teams will be required to maintain and test every pipeline component that is built to ensure consistency.*

Those from the DevOps team that were tasked with running proof of concepts for different tools were finished. They had brought a list of the tools tested, their benefits, and also their drawbacks. The full platform, regardless of how many tools were needed, should be able to fulfill all of the CI/CD requirements for their company. One major area of concern and importance was cost of ownership, which would either be listed as a drawback or a benefit. Open source software would be the preferred option since there was no initial cost to the company. However, because open source software does not come with support, the team would also need to add third party support costs into the evaluation. Any licensing and hosting costs would also need to be considered, as well as expected ongoing administration requirements. With all of these considerations in mind, the teams begin their deliberation of all the tested solutions.

The most popular solution that everyone on the team is the most familiar with by name is Jenkins. This solution is one of the most popular solutions in the software industry since its release in 2011. The proof of concept for Jenkins started with the teams taking a crash course in Groovy, since that was one of the main languages that would be leveraged by the tool. Then, based on the requirements gathered by the different interviews, the team downloaded and set up Maven and then looked into the Jenkins plugin marketplace for required plugins. Plugins are community authored and maintained code bits that can be added to a Jenkins instance for advanced support and feature enhancement. To start the proof of concept, the DevOps team installed a few of the more common plugins, such as Credentials, Kubernetes, job **Domain Specific Language (DSL)**, *and the pipeline plugins. Although more plugins may be required, these were the ones that the team initially started with. Additionally, the team had to become very familiar with the master-slave concepts of Jenkins and how those were to run and scale, especially when multiple cloud regions were to be used.*

Because of how Jenkins is designed, there were a lot of native capabilities for building the artifacts, but everything beyond that step required Groovy and Bash coding. These requirements in Jenkins made for a more imperative delivery and deployment process. The team considered leveraging shared template libraries and Jenkinsfiles to make the pipelines declarative. However, Jenkinsfiles required significantly more information than just application configurations. And training future DevOps members on how Jenkins is setup to be declarative would be difficult.

By the end of the Jenkins evaluation process, the DevOps members decided that Jenkins would suffice for their integration requirements, but the delivery and deployment requirements would need to be handled by other tools.

Other open source solutions that were evaluated were Drone CI and Travis CI, both being among the highest-rated CI solutions on GitHub. The installation for each was very quick, with Drone being the simplest and smallest of the installation and configuration requirements. In fact, compared to Jenkins, Drone was about a tenth of the size requirements and the plugin library was less convoluted and had easier custom plugin capabilities. The main issue with Travis CI is that it only works with GitHub, which means that the company would have to select GitHub as the source code manager to leverage it. But, since the company has not made an official decision on GitHub or BitBucket, Travis CI was still in consideration. Travis CI has some better parallelism built in natively, which would allow shorter build times, as well as centralized reporting for management teams.

The team evaluated some vendor-based solutions as well, namely CircleCI and GitLab CI, which provided significantly more power and advanced feature capabilities than what the team had previously considered. Some of the major benefits were features such as automated build types based on language detection, easy setup leveraging declarative configuration files, and significant native integrations with other tools.

Once the integration tools were presented and considered, the teams moved on to delivery and deployment tools. The first tool evaluated as a platform-wide tool was Jenkins, but the process of building out all of the delivery and deployment requirements was very time-consuming and the team had to present the solution as more hypothetical than a reality. Some of the open source solutions considered for the delivery and deployment pipelines were Spinnaker, by Netflix, and Argo CD which is a popular GitOps tool built by Applatix.

The process to get Spinnaker installed and set up was similar to Jenkins, in the sense that the solution is very heavy on hosting costs. The software required 10 or more microservices to be installed, although the install tool, called Halyard, did make that process much easier for them. Once the solution was installed, the heavy administration effort requirement became immediately apparent. Each delivery or deployment pipeline has to be imperatively defined and configured, for every artifact, for every environment, for every application. Additionally, since access control is limited to an entire application, user access would be difficult to enforce. However, Spinnaker did offer some great insight into how the infrastructure was behaving, with easy infrastructure maintenance capabilities, such as scale-down and scale-up commands. Spinnaker also offers a very robust API, allowing for significant integration and customization capabilities.

Argo CD was much easier than Spinnaker in regard to the setup. The install process started with a quick Kubernetes command that installed the Argo operators into the cluster with some minor follow-up steps afterwards. The team was able to quickly log into the Argo CD solution, connect to GitHub, and deploy. Argo CD also provided an up-to-date health status for the deployed services. However, that was all that Argo CD offered for the company. The installation was quick, and the deployment was automated when the Git repo was updated. But not all of the delivery requirements would be able to be handled by Argo CD just the deployment.

As for the vendor-based solutions, the main ones considered were GitLab, since GitLab CI was tested for the integration process, as well as Harness. GitLab has an entire platform that extends into source code management, monitoring, testing, and more. The advantage of the GitLab platform is that all of the pieces should work together very well since it is all the same company. However, the team had some pricing and vendor lock-in concerns that would limit GitLab to only being used for either CI or CD. GitLab CI was very impressive in the integration phase, and GitLab's Auto DevOps offering would allow support for delivery and deployment as well. GitLab Auto DevOps did offer some great code-based configuration capabilities, and even an automated pipeline suggestion feature based on the code repository that it is tied to. What the team found was that the pipeline suggestions, although useful, would require significant customizations to meet their needs.

Harness offered a very compelling delivery and deployment solution. The hosting costs were very low, since only a delegate needed to be installed in the network, and the administration effort was rather simple during the proof of concept. The templating features, the configuration code capabilities, and the large amount of native integrations allowed for easy connections to the other tools that the business already uses. As the team explored the tool more, they found that the configuration process, at least initially, would be very administrator heavy. Also, the tool did not offer any signficant reporting capabilities on the current services, unlike Argo CD or Spinnaker.

The DevOps team decided that they needed a more comprehensive examination of some of the tools. They decided that they would evaluate an open source and a vendor solution in each category. Each tool would have to show how easy it would be to build and scale a templated pipeline that had as many company requirements as possible.

Most organizations have to leverage a long evolution process of change to alter their habits or tools internally, especially when user adoption and culture have to be considered. Oftentimes, a tool can be exactly what the business needs, but because the users refuse to adopt it, the tool is discarded or shelved as a result. Most teams want to build a set of *happy paths* which can be leveraged by an entire engineering organization. Sometimes teams are presented with change agents that allow for more radical changes to take place. These change agents might be new leadership, a new paltform requirement, or even a new business initiative. But as most teams look to improve or change a process, they will quickly realize that replacing what already exists is difficult.

The first process in choosing a tool is not deciding which tools to use or implement but rather which tools are not going to work for the teams or processes. By defining the requirements and end goals at the outset, choosing the tools that will not work gives the team a better understanding of what they should be looking for. And, since the DevOps team will become the central CI/CD tool administrators, they will want to understand every in light of where the majority of the administration effort will lie: when the solution is implemented or as it is being used.

It is inevitable that every solution will require work to be done by power users. Some tools will require most of the administration effort upfront. But the trade off is that ongoing administration effort should be minimal. Other tools are very simple to get up and running, but the administration effort for the long term will be extremely taxing. And a third option is a tool that requires some level of effort consistently throughout the entirety of the solution's usage. it is difficult to understand how much work will be required and where the majority of the work will be. With open source solutions, the majority of the administration work is commonly frontloaded. The installation, configuration, and scaling of the solution requires significant effort. But once the tool is setup and working, the ongoing effort should be rather minimal. Vendor-based solutions are typically lighter in regard to initial effort and ongoing requirements, but the ease of use is offset with licensing costs.

Whenever a new solution is considered, it is paramount that the process is defined first. With regard to CI/CD practices, if a team understands what requirements exist for their company, then a solution must meet those requirements. It is rare to see a DevOps team understand what is required before engaging in building or buying a new tool.

Needs versus wants

It wasn't until the second proof-of-concept for these tools that the teams started to get more attention from the business leadership. The CDO was curious to see what the DevOps team was doing, especially if there were measurable effects of the recent DevOps initiatives. The DevOps team was able to share what requirements had been gathered, what tools were being evaluated, and so on. For some, the speed at which these steps were accomplished were extremely fast. But the CDO was not as impressed with the timeline.

The initial directive and timeline, which was the main driver for all of the work the DevOps team had done, was now being altered. The CDO needed the teams to have a beta version of the application deploying to Kubernetes within a month.

The DevOps team immediately went into panic mode. Although the developers would be able to get the artifacts ready in time, the delivery pipeline would not be. The amount of time available meant that the DevOps team would have to reduce the scope of the pipeline. Only the most basic requirements could be supported going forward.

As the future process was still hypothetical for the DevOps team, they had to understand what the initial pipeline outcome should address. One of the most basic needs for the teams was to automate the deployment of the artifact into a designated environment. with the needed configuration variables. The pipeline would also need to account for failure actions, whether that was a simple notification that would alert the DevOps team or some form of automated failure strategy. Lastly, the pipeline process would have to provide a solution that was easy for the internal teams to operate.

This new reduction of functionality immediately added a mountain of technical debt, before the pipeline was even built. They had no time to adequately evaluate which tools to use, nor did they have budget available to purchase a tool. The team needed a short term solution, even if they would have to tear it down and rebuild it later. The tools with the quickest time-to-value for deployments would be the immediate choices. The team decided to use Jenkins for the integration, since that was the most familiar and Argo CD for the deployment process.

All of the security, testing, and configuration requirements would have to be manually executed. The DevOps team knew that the solutions chosen would also add technical debt but the most important metric to monitor now was Mean Time to Executive Satisfaction.

The pressure that can come from a top-down technical or business initiative onto an engineering team is very heavy. This is mainly because the business doesn't actually understand the technical requirements to accomplish the end result. Their focus is more on the market, potential customers and that shareholders want a solution in the shortest time, with the best quality, and at the lowest cost.

Engineering teams are always tasked with building a solution that is low cost, high quality, and quick to market. These three requirements make up the project triangle that every initiative fits inside of. The only problem with the project triangle is that only two of the three can exist together. The most common requirement of the three is the time-based one, leaving leadership to decide on whether they want the solution fast or for less cost.

To achieve a desired timeline, it is common in the industry to navigate towards the more familiar tools. The time required to learn and understand is often not a luxury most teams can afford. Every solution balances capabilities with complexity. Anything that is more complex will also have more capabilities, and the opposite is also true. But a common trap that teams fall into is the assumption that leveraging multiple simplistic tools will be better than one complex tool. It is important to understand the entirety of the problem and desired outcome of the solution before pursuing a solution.

The automation test

As the 2-week deadline was quickly approaching, the selected tools were being installed and configured to accomplish the needs of the initial design. The DevOps teams set the deployment pipeline to when a GitHub repository was updated. They also had to create each Jenkins build and link them together to create the desired build pipeline. The Jenkins pipeline would create and upload a Docker image with a "latest" tag to make the integration process move faster.

Although the building and uploading of the Docker image to the container registry was a relatively easy task for the team to script out, they had to figure out how to deploy the container to Kubernetes. Executing the Kubernetes command with a manifest was not too difficult to run from Jenkins, but the issue was generating the YAML manifest. The team decided that the fastest solution was to leverage a copy of a public Helm Chart. They would customize it to meet their needs and give the developers limited override capabilities. After Jenkins generated the Helm Chart a job would run that would upload the Helm chart to a Git repository, triggering Argo CD to automatically deploy the Helm Chart. This first phase of the pipeline was executing consistently and without error.

The main issue that the DevOps team faced now was that every step after the initial deployment was manual. Although the new version of the Docker image was being deployed to the development cluster automatically, the developer had to manually test the new artifact. Assuming that the manual tests passed, the developer would then have to log into Jenkins and manually execute a different Jenkins Pipeline . This new pipeline would send the updated YAML file to a different GitHub branch, which would trigger Argo CD to deploy to the QA cluster.

The DevOps team demonstrated the deployment pipeline to the management and engineering team, which garnered mixed reviews. The management team was excited that the artifact was automatically deployed. But the engineering team was only focused on all of the manual requirements. This pipeline could not be the main process going forward. The DevOps team needed to figure out a better solution.

With the old process still needing support and the new pipeline being used, the DevOps had significantly less time to build something new. They had contemplated building out the Jenkins portion to supplement the different requirements. However, that would only increase their technical debt with additional maintenance. Even with the little time available, the team decided to look into some of the other tools they tested to see if there was a better option.

Automation is only as good as the resulting burden. Stepping over a dollar to pick up a dime is not worth the effort in the long run, and so it is with the software delivery process and the desire to automate everything. Many teams look to automate a part of their process, but they often don't realize the manual burden they created. In many cases, the manual burden is related to the maintenance of the automation capability needs.

Every company in the industry that has to work with software delivery consistently faces the product development triangle: short time, high quality, and low cost, and they can only pick two. Yet, as teams grow in their automation capabilities, mean-time-to-market, and as companies gain more market share, the product development triangle shifts to a product development square, adding low technical debt to the equation without altering the available outcome choices: you only get two.

Summary

The goal of this chapter was to build upon the foundational concepts from the last chapter and discuss how different companies in the industry solve the delivery and deployment process internally. A team wanting to implement continuous processes related to delivery and deployment should not start by choosing the tools first, but rather by understanding what the outcome of the process should be.

Every engineering team, especially a DevOps team, will encounter difficult delivery timelines and will need to evaluate what true requirements are and how to best accommodate the desired outcome with the desired requirements.

The next chapter will cover GitOps at a high level, some considerations that a team should have when implementing GitOps, and some of the nuances that can make GitOps adoption difficult.

3
The "What" and "Why" of GitOps

The last two chapters covered the basics of delivery and deployment. But, more importantly, they provided a brief overview of how most teams in the industry used to operate and why they moved to a continuous practice.

This chapter covers why delivering and deploying continuously is so important. Additionally, the chapter will cover how teams use declarative process to achieve a desired deplyment frequency while reducing risk.

This chapter covers the following main topics:

- Why does a continuous process matter?
- What is GitOps?
- Declarative execution and declarative state
- To push it or pull it ?

Why does a continuous process matter?

As the DevOps team continues to support the old and new release pipelines, the technical debt continues to grow. With every service that is onboarded, the DevOps team becomes more of a bottleneck for developers.

The original goal of the pipeline was to be able to release as often as the developers would want. But with the major restructuring of the delivery and deployment because of the business requirements, the quality and development teams were getting in each other's way. The new development process consisted of a developer triggering a code merge with the main source code branch whenever a new piece of code was written. Since every code change was resulting in the integration process being triggered, the development team was producing release candidate artifacts faster than the quality team could check. This resulted in a large set of artifacts waiting to be deployed into the testing environment. The quality team had to reduce test coverage in order to keep the pipeline flowing.

However, since the quality team was unable to pause the pipeline before dozens of different artifact versions became stuck in the process. Every engineering team was negatively impacted by the issue, and they all let the DevOps team know. The impact range of the issue started to remind the DevOps team of the old quarterly releases. The goal of the pipeline was to reduce issues and manual interventions, but that was not reality.

The DevOps team needed to work through a Root Cause Analysis process to understand what went wrong and why. Yet the RCA could only be accomplished after the issue was resolved. The team needed to figure out a way to remove the current blockage and prevent the issue from happening again.

The pipeline process was immediately paused between the artifact repository and the pipeline trigger, preventing new artifact versions from being submitted to the pipeline queue. The quality team was instructed to track and reject all queued artifact versions. This would send the rejected artifact versions out of the pipeline, operating as a makeshift escape hatch. Once the queue of artifacts was drained completely, the quality team would take the most recent version of each artifact, manually pull that artifact version from the artifact repository, test the artifact, and send release candidates to a pipeline that was cloned by the DevOps team.

Meanwhile, the DevOps team decided that having the different development teams reduce the frequency of executing the integration process would prevent a flood of artifacts from breaking the pipeline. The frequency reduction meant that the teams would only push larger sets of code changes. The team leads would be notified when a new artifact was built. Only the team leads where able to deploy, which had to be executed manually for a time. The deployment frequency dropped from daily deployments to weekly deployments, which lightened the load on the pipeline.

Within a few weeks of fixing the queueing issue, teams started to experience a different type of problem that they had not seen since their shift to microservices. Development teams were producing entire features for every build, but the new features depended heavily on other services that were also being changed. The teams started to realize that the many different services were now tightly-coupled to each other. And not only where the services tightly-coupled, but service versions were tightly-coupled too. This meant that teams could only develop code on a feature-specific Git repository branch. These long-living feature branches had many adverse affects for the development teams, such as poor test coverage and merge hell. The developers had to keep a manifest that mapped the different versions of the services that had to be deployed together. The version-mapping manifest was helpful for internal testing. However, the manifest was rapidly growing in size, and also became a requirement to mitigate failures. This meant that the version-mapping manifest had to be versioned itself. Eventually the DevOps team was needing to package all services together from the version-mapping manifest to prevent incompatibility during deployments.

What started as a quick-fix solution to relieve a bottleneck and allow teams to get back to daily deployments resulted in **Service-Oriented Architecture (SOA)**. *SOA is a development practice consisting of microservices that have the same tight coupling of a monolith. The SOA development style will often result in weekly or monthly deployment cadences. If the DevOps team could not get the process under control, the desire to achieve continuous delivery and deployment would be impossible.*

Anytime that a centralized team, such as the DevOps team in this analogy, owns the delivery pipeline process, the other teams that are supported by the process become customers of the centralized team. The benefit of the centralized team topology for delivery and deployment is that standardization is built into the entirety of the process. However, the drawback of the centralized team topology is that when something in the process breaks, it can result in a massive bottleneck for all of the teams.

The engineering organization in the analogy experienced a very common issue when a continuous model is first adopted; increased deployment frequency outpaces manual execution capacity. What the DevOps team encountered was the successful deployment frequency from the developers into the quality environment. However, the quality teams had not yet built out a full testing automation process. This meant that the DevOps team was not able to add the automated trigger for the delivery pipeline to deploy from one stage to the next. Instead, the quality teams had to manually test, manually document, reject or approve, and execute the next step in the delivery process. This inevitably led to a bottleneck in the testing queue. Artifact versions were being developed at a rate that the quality team could not match. Often, this bottleneck will lead to brittle deployments because the quality team is pressed to relieve the artifact queue rather than improve testing. The business will initially become more concerned with meeting customer requests rather than platform stability. Ultimately, the blowback from leadership resulting from poor production quality will lead to significant engineer attrition and customer churn. In the case of the engineering organization in the analogy, the business was mandating a deadline that each team desired to meet. But three independent teams working independently will consistently encounter communication issues, especially when a deadline is hanging over their heads.

Business-mandated deadlines on an engineering project are often poorly calculated. And as deadlines get closer, teams are tempted to cut corners to succeed. Security becomes an afterthought, as does monitoring and alerting. Break-glass alternatives are not properly engineered or implemented. And when the product fails, developers are forced to find any workaround possible.

Many DevOps teams find themselves operating more as firefighters, rather than DevOps practitioners. A rushed team will build a low quality or incomplete product. The result is a constant state of hotfixes and short-term workarounds. A common buisness mandate in the current age is containerization and cloud migration. Although both of these initiatives can be beneficial, if they are implemented poorly they will result in massive techincal debt.

In a monolith, all interactions with different features and functions are built-in. When a company wants to adopt microservices, one of engineering's first hurdles is figuring out how to get the services to talk to each other. Some other signficant differences that need to be considered are database interactions and backwards compatibility between services. One common solution is to leverage a centralized staging environment for interactivity testing. Every service will have it's most recent production-ready version deployed there. Then, as a team needs to test a new version, they will deploy it, test it, and then revert the versions if needed. At that point, the service version is approved by a team lead and the version number is added to a release manifest. A general count of all services in the centralized environment is recordedto make sure that all services are accounted for. Each service will have a different rate of development, leading to different versions for every service in the environment. Once each service has a release version ready in the environment, the team leads will approve the manifest for deployment. A release captain will take the finalized release manifest and set a time to test the full deployment into a near-production environment. A successful test deployment will result in a production release date. This process pattern is common when a company is leveraging a Service Oriented Architecture (SOA) for their application. Most companies that want to move from monoliths to microservices will typically find themselves in the SOA pattern, and many rarely make it out.

There are some obvious issues with the release process mentioned earlier, mainly that development is restricted to the rate at which the slowest team operates. However, some other issues with the release process mentioned are stagnant code, reduced developer productivity, increased deployment failure risk, and difficult rollback capabilities. Whenever services are developed in a tightly coupled way , deployments become infrequent. Teams that have higher rates of productivity will typically be held back by teams with lower rates. Although SOA is not necessarily a bad application architecture model, infrequent deployments are hard to avoid. When an application is deployed infrequently, the code changes become larger. Large code changes result in a higher probability of bugs. More bugs in production mean that more customers will be affected. And when customers are affected, confidence in the team, application, and company begins to drop.

Delivering and deploying in a continuous manner reduces the number and size of changes. Smaller changes reduce the number of potential bugs in the code, which decreases the time to deploy. Smaller changes also lead to easy rollbacks , which result in a high degree of confidence from everyone involved. When the delivery and deployment processes are imperatively defined, they become brittle. When teams are required to leverage brittle processes, they will often implement their own monitoring and troubleshooting practices. By leveraging a declarative model for deployments, the solution will naturally be more scalable. And when the behavior of the solution is well defined, then reliability is easier to achieve. The ability to deploy continuously is almost entirely predicated upon how reliable and repeatable the platform is.

What is GitOps?

The DevOps team found themselves in the middle of a perfect storm of problems. They were still needing to support the quarterly release process, at least until the monolithic application was completely broken down into microservices. They also had to support the new pipeline process that they had builtwhich has resulted in major support issues. And the team also needed to figure out a new pipeline process that would eventually supercede the rest.

The hotfix for the pipeline queuing issue was now completely resolved and the teams could move back into the previous continuous process that they had before. However, since the development process had become a tightly coupled service oriented architecture, it was difficult to get the teams to adopt anything new. Ultimately, the DevOps team needed to not only bring in a better process, but they also had to make it easy to adopt, automate, and reduce the onboarding requirements.

Since the developers were already familiar with Git repositories, the DevOps team wanted to start there.

By leveraging the Git repository that the development teams were already accustomed to, the delivery and deployment process would actually be significantly easier to adopt. The Kubernetes manifests and application codewere already stored in Git. And the teams were well accustomed to Git versioning and branching. Also, since the teams were using pull requests to trigger the builds, the DevOps team could continue using the same triggering event.

The DevOps team also found that the cloud teams were storing their Infrastructure-as-Code configuration files in their Git repositories as well. This allowed for the potential of a larger scope of support in the desired solution. By leveraging the Git repositories as a central source of the truth, the DevOps team could adopt a GitOps process.

GitOps has been a rather popular topic in the DevOps industry recently. The DevOps team was looking for a set of best practices on using the Git repositories more. And when they researched deployments and GitHub, GitOps was the main result. Because of the recent pipeline issues, the DevOps team wanted to approach the new pipeline differently. GitOps sounds like a great concept, but rushing to implement a tool would result in the same type of outcome as before. The team needed to understand the process and find a tool to fit.

Conceptually, GitOps is the practice of storing declarative state files in a Git repository, which declares the desired state of the application in a designated environment. In practice, GitOps has been siloed to Kubernetes and Infrastructure-as-Code because the underlying platforms rely on declarative files for their configurations. When the team considered the platforms that the application was built on, they realized that the configuration requirements would be minimal. And since the GitOps tools required little user interaction, the DevOps team could implement the tool without interrupting the developers. The only piece that the DevOps team had left to consider was how to avoid the artifact bottleneck issue they encountered before.

Reactive engineering, or the process of improving a product or application solely based on responding to the feature enhancement requests of users, is a very common style of product development. In fact, most companies design their applications based on the immediate needs of the market and rarely engage in improvements outside of the reactive cycle. This process often leads to developing the minimal functional requirements to release the product with the shortest time to value. The ultimate outcome of this reactive engineering process is significant technical debt.

With support engineering, where the sole focus of the support team is to build and maintain the support system for the developers, the issue of reactive engineering is a significantly larger problem. For DevOps teams, which support the CI/CD processes, the integration, delivery, and deployment practices already exist in some form or fashion. This means that any changes to the process is similar to building the car while driving the car.

Many DevOps teams would like to build a new system from the ground floor. Being able to adopt the newest technology and practices to support the engineering organization is DevOps nirvana. Yet many DevOps teams encounter major resistance when even small changes are proposed. As a result, many important changes happen when no one is looking. The best way to bring in something new is to make it look or operate like what already exists.

GitOps has been practiced in some form by many engineering teams since the beginning of the declarative configuration management movement with tools such as Puppet, Chef, and Ansible. The term was originally coined by Weaveworks in 2017 to define how they leveraged Git as the source of truth for their application and cluster definitions. Prior to GitOps being a formalized term, most companies would leverage configuration files that would alter the behavior or outcome of a deployment or task.

Many engineering teams prefered the declarative style of the configuration files. The centralized administration team would be able to build guardrails into the execution engine. Because of this, other engineers could add whatever they wanted to their configuration files, without hesitency. GitOps is the logical extension of this process. State files are stored in a Git repository with an event, such as a pull request, triggering the desired change to take effect.

Since most teams are already familiar with Git best practices, implementing GitOps was relatively simple. Where some engineering teams might see issues is with the fact that GitOps tools, up to this point, have been almost entirely focused on Kubernetes and Infrastructure-as-Code.

Declarative execution and declarative state

To start the groundwork for the GitOps process, the teams had to understand what repository structure the different teams were using. GitOps tools tie individual Git repository branches to specific environments. Since the developers are deploying to three or four environments, the GitOps tool of choice would need to be configured to support these environments. Adding to this requirement, the deployment pipeline needed to sequentially deploy to the different environments. The DevOps teams would need to figure out how best to trigger the deployments automatically.

The next major step was to get the correct pull request approval process to be executed. This would allow the Kubernetes manifests in one branch to merge with the Kubernetes manifests in another branch to trigger the deployment. The linking of the tool in the right cluster to the desired repository branch would need to be automated leveraging the GitOps tooling API, effectively reducing the amount of manual work by a significant margin. The tool that had previously been evaluated and best fits their immediate needs would be Argo CD. Argo CD is a GitOps tool that lives within a Kubernetes cluster and enforces the workload in the cluster to match the manifest in the repository. This means that Argo CD treats the Git repository as the single source-of-truth.

The DevOps team had not encountered the concept of a desired state before exploring GitOps. And since Kubernetes manifests are essentially codified desired state files, the concept was not too difficult to understand. After all, every other deployment pipeline being used was accomplishing the same thing. The user would define what they desired the end state to be, and the pipeline would work to accomplish that. However, declaring the desired end state of the Kubernetes service was completely different than declaring the desired execution process as a whole.

A few months prior to the hotfix fiasco, the team had come to an understanding of the differences between deployment and delivery. Deployments were simply process executions that focused solely on getting an artifact into a desired environment. Whereas a delivery contained all of the pre-deployment and post-deployment as well. What Argo CD focused on was the deployment process through GitOps. However, if the team wanted to automate the delivery process as well, a tool would need a declarative state file and a declarative execution file. This would allow the DevOps team to declaratively define the desired end state and desired execution state of a delivery and deployment.

The timeframe for the implementation of a new GitOps process was very short. The DevOps team realized that they would need to hold off on the desired delivery pipeline and focus only on the deployment process. Significant manual intervention would still be required to avoid bottlenecks, but there was no other way to meet the deadline.

To start testing GitOps deployments, the DevOps team had to configure a few things. They need to get a fresh Kubernetes cluster for testing. Then they needed to get Argo CD installed and configured in the cluster. Finally, the team would want to test the automation by connecting Argo CD to a real Git repository. Once these steps were complete, Argo CD should be able to start deploying automatically on the next pull request. The next pull request to the Git repository triggered the deployment process, as expected. Argo CD saw the new manifest, ran the appropriate commands, and reported back a healthy system. When the deployment was complete, the DevOps team saw a new chat message from Argo CD confirming the deployment status. But as the day went on, no new deployments were being triggered. Developers had been pushing new code, and new artifacts were available, but Argo CD was not deploying as expected. What the DevOps team failed to realize is that the manifests were not being updated with every artifact build. Because the teams were using "latest" as the artifact tag, they had no need to change the manifest. And with the manifest not being updated, Argo CD was not being triggered.

Another major issue with using and reusing the latest *tag meant that no one actually knew which version of the container was being used. It was impossible for the DevOps team to know which one was the "latest" and which one was the "latest latest." Additionally, since every new version was being uploaded as* latest, *the upcoming auditing cycle would require significantly more effort and time.*

Instead of leveraging semantic versioning, which is the use of major, minor, and subversions (that is, 1.0.0, 1.0.1, and so on), the developers were using container hashes. The version mapping manifest would include the required artifact hashes. But the deployment pipeline would only delete the Kubernetes resource, pulling down the new artifact as a result. As a result, the DevOps team needed to figure out a better solution. They could require the developers to adopt versioned tags on the artifact, but that would be a significant change. Another option would be to manually trigger Argo CD to rerun the deployment. One last option would be to require the teams to add an inconsiquential change to the manifest to trigger Argo CD. But if the deployment trigger was to be based on a synthetic change in the manifest just to trigger the deployment, the purpose of using a tool such as Argo CD would no longer exist. Although the team needed to figure out the best path forward, they started to wonder if pulling the artifact down was ideal. Since Argo CD lived inside of the cluster, it would essentially be pulling down the manifest and artifact. This meant that if the Argo CD cluster went down, no other deployments could happen. It might be better to find a tool that lived outside of a cluster and could push the changes instead.

The two biggest hindrances to adopting a new process or tool are the culture of the team and user adoption. If the users are willing to adopt the new process or solution, but the culture of the team prevents best practices from being adopted, the tool will fail. However, if the culture of the team allows for best practices to be leveraged, but no users adopt the solution, the tool will fail.

One of the major advantages of GitOps is that it leverages tools that are already in use by the development teams. And since the execution is based on an update to the Git repository, the deployment can happen at their desired development frequency. Most GitOps tools are limited in their configurability, resulting in limited alternatives to the tools best practices. Because of the ease of use and limited configurability, GitOps tools often succeed, regardless of user adoption or team culture.

Kubernetes-based GitOps tools have another advantage, which is Kubernetes itself. GitOps tools leverage Kubernetes manifests to define the desired state of the workload. Kubernetes will actually act as the execution engine, because of how Kubernetes is constructed to operate. If a user wants to implement GitOps within a Kubernetes cluster, it is important to understand the basics of how Kubernetes works. Kubernetes is made up of many components that work together to achieve the desired outcome for a cluster. `kube-api-server` is the central interaction point for the Kubernetes cluster and handles all of the requests coming from the `kubectl` commands and manifest files. `kube-api-server` talks to all of the other components and interfaces with the central storage component called `etcd`. This storage component keeps a record of all the parts and pieces of the Kubernetes cluster, allowing a consistent and up-to-date record of the cluster. When a command is executed and validated through `kube-api-server` to provision new resources, such as a deployment, another set of components is activated to fulfill the request: `kube-scheduler`, `kube-proxy`, and `kubelet`. These components work together to understand what resources are available on which underlying node and how the resources from the creation request will be provisioned. To add to this behavior of creation, `kubelet` interacts with `etcd` and `kube-api-server` to report back what hardware resources the node has and what Kubernetes resources exist on the node. As a result of this interaction, `etcd` understands what the current state, or current reality, of a Kubernetes resource is and what the desired state, or desired reality, of a Kubernetes resource should be. For example, if a command tells Kubernetes that there should be a deployment with three replicas, the command tells Kubernetes the desired state and Kubernetes attempts to achieve the desired state if it is different from the current state.

The term declarative state has been defined already, but what most processes are missing are the declarative execution files. Similar to the declarative state files containing the desired end state of a particular application or infrastructure, the declarative execution files contain the desired execution behavior to achieve a desired application state. When considering a platform such as Kubernetes, there are few execution file requirements involved since Kubernetes handles the execution for you. The closest a Kubernetes administrator could get to the process would be either running Kubernetes jobs, using an affinity or probes, or leveraging an advanced admission controller process similar to Open Policy Agent.

To push or pull ?

As the DevOps team considered the current versioning issue and updating the Kubernetes manifest files, they also wondered whether the GitOps tool should be inside or outside of a cluster.

Some of the benefits that the team came up with were things like direct cluster access. This was especially helpful with the monitoring and enforcement of Kubernetes resources. The tool would also have immediate knowledge of the health of the resources. Another benefit is that the execution engine should be fairly small with a low cost of ownership. Lastly, since the execution happens within the cluster, all data and secrets reside in the cluster as well. This would prevent the need for consistent polling for information from an external source.

There were also some benefits of having a GitOps execution engine live outside of the cluster. Most importantly, the execution engine is not dependent upon the life and health of the cluster. If the cluster, or the node that the execution engine lived on, was to go down, the entire execution engine would be wiped out. Spinning up a new node or cluster and re-adding the execution engine with all of the desired connections would not be a quick process. An external execution engine could also deploy to all clusters from a single location. Another major benefit of an external tool would be the ease of configuration and restoration if a cluster or node goes down, since the health of the tool is not based on the cluster being in a healthy state.

Argo CD offers a helpful user interface to leverage when wanting to connect different repositories and allowing teams to use different templating languages for the Kubernetes manifests. But the tool is entirely dependent upon a cluster being healthy and has a very difficult time handling deployments of applications across multiple clusters.

The DevOps team needed to decide on which features were most important. An external tool would offer ease of observability and configuration across multiple clusters. But a such as Argo CD provided a quicker time-to-value and lower cost-of-ownership.

The debate about whether deployments should be push or pull-based has raged on for a few years. This argument is almost always coming from companies that offer a solution that either leverages a push or pull-based method. The reality is that the best style is relative to the team. Some teams want an external solution and rely on network connectivity. Other teams might have network requirements that make pull-based methods more beneficial.

When the application architecture consists of virtual machines and servers, the push deployment method has a significantly lower cost of ownership because a central location pushes the deployment to all endpoints. Whereas a pull-based deployment method would require the teams to have an agent living on every endpoint to deploy. In a pull-based deployment, the agent will have all of the internal mechanisms to make the connection between the central server and the agent easier to manage. But a push-based deployment method, a central server requires security keys or certificates and service accounts on every endpoint, as well as any network access.

With an architecture such as Kubernetes, the pull-based deployment method is significantly easier to administrate since permissions and network connectivity are typically configured for the cluster already. And since common practice with Kubernetes is to only have a handful of clusters, the cost of ownership is significantly lower. For push-based deployments, an external deployment engine would require additional security and network configurations. Push-based deployments to Kubernetes will also require a constant polling mechanism to check for health and status. But, one major advantage to push-based methods for Kubernetes is the ability for one central engine to interact with multiple clusters. This allows for easier admission control and data correlation across multiple clusters.

If a company wants to understand whether a push or pull-based deployment method is better for their application, they need to understand the benefits of each option, . The most important pieces to consider are the network and security requirements for the company. These two areas will often dictate which method will be the easiest to adopt.

Summary

This chapter finished the overview related to software delivery, deployment, and the importance of continuous processes. New concepts were also introduced related to GitOps, how GitOps is currently practiced in the industry, the difference between declarative executions and declarative state, and a high-level overview of push-based and pull-based deployments.

The goal for this chapter was to provide enough information for you to understand what GitOps is, common practices for GitOps, and some of the nuances in implementing GitOps that a team should consider as they begin their adoption of the GitOps processes.

Section 2: GitOps Types, Benefits, and Drawbacks

In the previous section, the fundamentals were covered and the foundation was set. Now, in this section, we will learn what different GitOps practices exist, how to choose the right one, how to define best practices, and where GitOps fits in.

This section comrpises the following chapters:

4
The Original GitOps – Continuous Deployment in Kubernetes

Understanding how the industry has arrived at the current state of continuous delivery and continuous deployment is essential in understanding the fundamentals of GitOps, what GitOps aims to solve, and why GitOps is important.

Even though GitOps is in its infancy, the practice has already evolved into a few different schools of thought. The first of these schools can be summed up as *originalist*, since it covers the practice expounded by the group that first coined the term GitOps.

In this chapter, we're going to cover the following main topics:

- Original GitOps basics

- Kubernetes and operators

- Manifest explosion

- Benefits and drawbacks of originalist GitOps

- Common originalist GitOps tools

Original GitOps basics

The DevOps team was now in a mad scramble to get a GitOps process implemented and adopted. The development teams had continued to break down the old monolithic application into containers to deploy into a Kubernetes cluster, resulting in significant growth of deployable services across all clusters and teams. The haphazard service-oriented architecture style of deployments with tightly coupled version dependencies would soon result in a deployment process that is too heavy to support. This would require significant team growth and manual intervention just to achieve the business objectives.

By implementing Argo CD in the cluster, the team was able to quickly have deployments automatically sync with the cluster tied to the tool. Argo CD is a GitOps tool and closely aligns itself with the group that first coined the term GitOps in 2017. The original intent is to allow an in-cluster engine to execute the desired deployment, known as auto-syncing, with the Kubernetes cluster that the engine is tied to. The two main areas that the original GitOps method is most associated with are Kubernetes manifests, like `Helm`, *and Infrastructure-as-Code files, like Terraform.*

When there is a change in the code repository that houses either the Kubernetes manifest or the Terraform files, ArgoCD will auto-sync the changes with the cluster.

Since the team currently leverage Terraform and Kubernetes, the originalist style of GitOps would offer a quick solution for the DevOps team. The only thing that the team needed to better understand was how ArgoCD worked in Kubernetes in order to not be locked into that tool in the future.

Deadlines and compelling events always seem to loom overhead as teams attempt to accomplish a desired outcome. The need to make a deployment process scalable and reliable is difficult to achieve when under a time-crunch. This desired outcome is precisely many companies in the industryare wanting to move to a GitOps model of deployment. The native tracking, versioning, and visibility that comes with storing code files in a Git repository makes onboarding of users and services easy. However, the process of adopting originalist GitOps means that the execution engine must have already built and tested manifests.

Prior to GitOps being introduced a few years ago, Kubernetes deployments required advanced Kubernetes knowledge. Any solution that allowed all users to have easy access to interact with Kubernetes was sorely needed. The only way to make that happen was for a tool to be an intermediary between the user and the cluster. Most teams looking to solve the problem had to build out an internal platform. But even then, most of the Kubernetes interaction was still delegated to an administrator group.

What the originalist GitOps practice capitalized on was the native ability of Kubernetes to handle the deployment and configuration of desired resources. At first, the users would create a sprawling mess of Kubernetes flat files, which was eventually improved through templating engines. In essence, templating engines allow for manifest files to scale through variable overrides. Because Kubernetes allows the declarative files to dictate what is deployed, the originalist GitOps process only had to facilitate the execution of the deployment. This means that the GitOps tool wasn't technically doing the deployment but rather just executing the deployment command.

One major benefit with originalist GitOps tools is the ability for the tool to understand and enforce the desired end state of a Kubernetes workload based on what is in the Git repository. This process of synchronizing the workload in the cluster with the manifest in the Git repository is not a unique concept. Kubernetes leverages the same synchronization design between etcd, the replication controller, and the kubelet.

One of the major issues with any form of automation that is provided by a third-party tool is understanding how that tool works. When leveraging a tool like ArgoCD to handle any GitOps process inside of a Kubernetes cluster, it is important to understand what ArgoCD is doing in the cluster and how. Being locked into a tool is not something that any team wants and understanding the underlying behavior will help to avoid the lock-in dilemma.

Kubernetes and operators

One major issue with GitOps, as the DevOps team was experiencing, is that the platform that the application relies on is the main execution engine for the GitOps process. Although the execution in the underlying platform being inherent is helpful, it also means that the execution is like a black box where the underlying actions are relatively hidden.

To avoid this black box issue with the GitOps process, the DevOps team decided to dive deeper into what Argo CD is doing during an execution event. By documenting Argo CD's underlying tasks, the troubleshooting process will become much simpler. The first step to understand and document is how Kubernetes behaves when an execution command is triggered.

Up to this point, the teams are basically just running helm commands. The team will monitor the helm output for a success or failure message, and then moves on. What the DevOps team wanted to understand was how Kubernetes behaved when the command was run.

The deployment process for Kubernetes applications is significantly different than the quarterly release pipeline. In the quarterly releases, the teams had to be very specific as to what needs to be installed on which servers, along with any configuration requirements. But with Kubernetes, the entire deployment process is inherent to the platform, with a manifest declaring the desired end state.

A helm *command would start the process by executing the desired templating command to bring the files together into a single large manifest. Then, the underlying engine in* helm *would execute a* kubectl *command to pass the manifest to Kubernetes. The* kubectl *command would pass the manifest from the YAML file to the etcd store via the kube-api-server. Etcd was then storing the desired end state of the application based on the manifest and other core system components would run certain actions based on the change in etcd. A controller would notice a change in etcd and move to steer the cluster to be in the desired state, which would involve the kube-scheduler, the kubelet, and other controllers. Once the resources were deployed onto the appropriate nodes, the response from the kube-api-server is a success message saying that the command was successful. When the resources were added to the appropriate nodes, the components on the nodes would then work to get the resources into the appropriate state by downloading the required containers and any relevant files. If there was an issue with the deployment, the users could execute a* helm *command to roll back the execution to a previous version and the previous process would execute again.*

This underlying process remained entirely unchanged when using Argo CD. The helm *process and* kubectl *process remained the same, as did the provision of the manifest. What the team was able to see from the Argo CD side is the use of a Kubernetes operator that would trigger and track the deployments instead.*

A Kubernetes operator is essentially another controller that is programmed to execute a desired process, such as maintaining a customized resource according to that custom resource's definition requirements. The operator would live in a state of constant checking to ensure that the custom resource was always up to date, which is the main concept behind the Argo CD operator. Argo CD considers the custom resource as the actual resource to deploy and the definition is the manifest stored in the Git repository. When the custom resource is not in sync with the Git repository, the operator, similar to a controller, will work to steer the deployment in the right direction. This is accomplished by executing the appropriate `helm` *or* `kubectl` *command on the underside.*

Another benefit for the Argo CD operator is that it can also treat the cluster as a custom resource and leverage an Infrastructure-as-Code repository as the resource definition to stay consistent. By Argo CD maintaining both the Kubernetes resources and infrastructure resources through the operator configuration the DevOps team should only need to concern themselves with getting the Git repository into the desired state and setting up the connection between Argo CD and the Git repository. The main parts for the DevOps team to document are the rollback process and what the manifest administration requirement will look like.

Kubernetes seems simple from a cursory glance at the execution requirements, since the process typically leverages a simple command and some manifest files. However, when considering what the Kubernetes platform is made up of and how it operates, many would-be Kubernetes practitioners can become quickly overwhelmed.

Harkening back to the first few chapters, the foundational concepts of an application instance living on a server and needing to be updated have not changed in the world of containers and Kubernetes. The two major differences are the size of the application instances and the container orchestration process.

Before Kubernetes, teams were orchestrating which servers to update how to update and so on. Kubernetes does this same thing, but with less hands-on administration requirement from multiple teams. Kubernetes leverages two core concepts inside of a cluster that make automated orchestration a possibility: etcd and controllers.

etcd, which is the in-cluster data store, is intended to keep a consistent record of the state of the entire cluster. When the user requests information of Kubernetes, either a request to create resources in Kubernetes or to retrieve resource information from Kubernetes, etcd is the source that the request goes to. The initial step of the create command is to write to etcd, which then trickles down the execution process to create the resources in the cluster, and then etcd reports back the outcome of the execution. The kube-api-server is the communication gateway for all users that want to interact with etcd, but also for every component in the cluster to interact with etcd as well. This is especially true with the other orchestration automation component.

Every Kubernetes cluster has a set of system components that help with cluster maintenance and administration. etcd and kube-api-server are some of these core components, but the components that actually execute most of the work in the cluster are the controllers.

Each controller in Kubernetes manages one or more resource types in the cluster. The management process consists of the controller reading what etcd says about that resource type, reading what the cluster shows relating to the resource type, and then working to make the two stay in sync with each other. Controllers will accomplish their tasks either by executing kube-api-server requests, or taking effects directly on the cluster. The complexity of the inner workings of Kubernetes are often uncovered the hard way, which is to say that they are uncovered during troubleshooting an issue. Therefore, it is highly recommended that anyone pursuing originalist GitOps practices should have a thorough working knowledge of Kubernetes first.

A popular originalist GitOps tool in the market is Argo CD, which leverages the operator model inside of the Kubernetes cluster. The goal of Argo CD is to make sure that the current state of a designated resource set inside of the cluster matches the desired end state that is declaratively defined in the Kubernetes manifests in a Git repository. This operator model with Argo CD is exactly the same as the controller model, but there is one significant difference: A controller treats etcd as the source of truth, while Argo CD treats the Git repository as the source of truth. The underlying behavior of Argo CD in relation to the Kubernetes cluster is essential to understand, especially when considering the administrative requirement for Kubernetes manifests and failure strategies.

Manifest explosion

The DevOps team continued their exploration of the underlying operations and requirements for GitOps with Argo CD and Kubernetes. One of the parts they also wanted to understand was the process of onboarding and maintaining the manifests for Argo CD to work correctly. The goal is to make a repeatable, reliable, and scalable process for any current and future applications across the grow cluster size and count.

One of the first things they noticed about the manifest requirements is that every microservice required its own helm *chart. Additionally, to add overrides files for different environments, the override files would have to reside in the same repository as the chart for Argo CD to leverage them. This Argo CD requirement meant that the DevOps team would have to maintain the core* helm *chart for every microservice and work through a pull request approval process for every chart or values file change on every microservice.*

Because of this, the DevOps team would have to come up with a core chart and approved override files that could then be shared out to every application team. The team would also need to automate the creation and linking process of the charts, the repository, and Argo CD in the appropriate cluster. But this presented yet another issue for the team to consider: how would they support one application or microservice that needs to be deployed across different clusters?

The quarterly release process had some major issues with the onboarding process because every step had to be specifically defined and changed ahead of time. The DevOps team had to provide a configuration file for every group of servers that had to be updated and the script required constant maintenance. Although ArgoCD offered the ability to get away from the issue of script maintenance and administration, an enterprise-wide adoption would be very difficult to adopt.

The only possible solution to make the administration requirements lighter for the DevOps team would be to move the `helm` *chart maintenance burden to the developers. It didn't matter who was responsible for maintaining the charts and files, the charts and files still need to be built and maintained. If the DevOps team did the building and maintaining processes, then they could enforce standardizations and easy compliance and auditing. But with developers are maintaining the charts, then different auditing steps would need to be in place.*

There are some who would argue that the role of a Kubernetes administrator is not really a Kubernetes administrator, but more of a YAML engineer. This is because the vast majority of Kubernetes configurations are handled through YAML files. Therefore, being able to understand how a YAML file should be constructed, what templating is available, and other Kubernetes nuances is required.

Prior to the templating abilities that came about through go-templating and `helm`, Kubernetes administrators would maintain individual Kubernetes flat files. There were tools that allowed for the user to provide some inputs that the tool would convert into a Kubernetes manifest, but the maintenance requirements still existed. Storing the YAML files in a Git repository was nothing new when originalist GitOps was marketed. Rather, the ability to leverage the Git repository as the application source of truth was the breakthrough.

As mentioned before, Argo CD, and other originalist GitOps tools, are essentially Kubernetes controllers that work to steer the cluster state towards what the git repository would show. This operator model is an extension of the current controller-etcd relationship within the cluster. However, the manifest requirements that originalist GitOps requires can be very cumbersome.

Something to note about controllers in Kubernetes is that it is very difficult, if not impossible, to have a controller in one cluster monitor designated resources in another cluster as well. The closest that a controller can get to this type of functionality is to increase the underlying node pool and leverage namespaces for everything. The reason for this impossibility is because the controller's purpose is to compare the current resource state in the cluster with the desired resource state represented in the source of truth: etcd.

Originalist GitOps tools have the same issue that controllers do. It is very difficult for the originalist GitOps tool to accurately monitor and steer the resources in the cluster toward the desired state defined in the git repository. Additionally, since the originalist GitOps tool is treating the git repository as the source of truth for a set of resources in a cluster, to spread a single set of manifests across multiple resources and multiple clusters is impossible.

Any in-cluster originalist GitOps operators will require a combination of manifest files for each microservice and each cluster. Adding to this manifest requirement, any overrides or templating that a user would want to have requires heavy setup requirements and maintenance for the originalist GitOps tool and designated microservice. For example, if a team needed different values to be applied for every environment that the `helm` chart is deployed to, `helm` can leverage a command flag to do this. However, to tell the originalist GitOps tool to accomplish the same outcome would require the administrator to execute setup commands when adding the desired repository link. With that command, the administrator would have to add any specific variables to override, similar to `helm` set flags, and values files to override, similar to `helm` file flags. To make the process even more nuanced, if the chart needed to be deployed to multiple environments with a different values file depending on the environment, the GitOps administrator would have to set up a separate deployment for the same chart on every environment.

The requirement for the manifest hierarchy process in the originalist GitOps configuration results in a manifest explosion that must be tracked, maintained, administrated, audited, and scaled.

Benefits and drawbacks of originalist GitOps

As the DevOps team dove deeper into the world of originalist GitOps and tools such as Argo CD, they discovered some very promising benefits of adopting this type of process. They found that Argo CD allowed for the mandatory Kubernetes manifests to be the main driver for configuration and change to microservice resources in the cluster. And since the developers were already used to using git, the adoption process of originalist GitOps automation would not be too much of a burden for the developers to work through.

By treating the git repository and helm *chart as the source of truth for the desired end state, the user should always be able to see what microservice configuration is in the cluster based on what the git repository shows. Also, since git has native versioning, a rollback to a previous version is as easy as reverting a version in git and letting Argo CD handle the deployment part.*

One other benefit that the team found was that if a cluster went down, since the helm *charts existed as the source of truth in the git repository, they could spin up another cluster and return it to the desired state in very little time. This shorter mean-time-to-restore for cluster recovery was a massive benefit for the company.*

There were some drawbacks that the team found as well. They had a set of helm *charts already created by the development teams, but the DevOps team didn't want to place the manifest creation and maintenance burden on the developers. Since Argo CD required that a chart existed in a single repo for every set of microservice resources that needed to be deployed to Kubernetes, there would be a manifest explosion.*

Another drawback, which is significant for the engineering organization, is that the developers have a different cluster for every environment, meaning that there are four different clusters that need to be deployed to. Unfortunately, Argo CD does not easily allow for cross-cluster deployments of the same helm *chart in sequential order. Adding to this issue, the company will require multiple production clusters, one for each major region of users across the globe. Argo CD does not have any way of being able to deploy the same chart across multiple clusters in different regions at the same time. The best option that the team could find was to leverage an Argo CD architecture known as an "app of apps," where a single git repository would store the files required to dictate applications and cluster, which requires manual executions only.*

The last drawback, and probably the most important drawback the team found, was that the originalist GitOps process is only for deployments and specifically for the easiest part of the process. Every microservice manifest had to manually be created and tested with every values file combination to every environment and cluster in the architecture. Then, when the files were finalized and added to the git repository, Argo CD could be connected to execute the install command for the microservice. But all of the other required steps, such as approvals, tickets, testing, verification, and others, would have to be done outside of the GitOps process.

Argo CD is a great GitOps solution and can solve many problems for different companies and teams in the industry. But the DevOps team needed to see if maybe there was a better GitOps tool that was native to Kubernetes and could handle all of their delivery and deployment requirements.

Kubernetes is not the easiest platform to set up, maintain, administrate, or scale. The ability to dive deeper into the inner workings of a Kubernetes cluster is available but can quickly become overwhelming. In many cases, a Kubernetes administrator wants to set up a Kubernetes cluster in the default manner and keep it that way. Kubernetes practitioners that build out YAML files for microservices deployments would also like to keep their manifests as simple as possible. But these desires are not the norm for the industry today. The growing security concerns around network, secrets, access control, and cloud platforms require signficant customizations. Some microservices require autoscaling, others require node affinity, and then there are microservices that rely on the cluster defaults for these types of specifics.

There is a massive advantage when leveraging a GitOps tool to handle the deployments of microservices. Because of the requirements for the microservices to have some set of manifests to dictate which Kubernetes resources should be deployed set up for the microservice to work correctly, the GitOps tool can simply leverage that requirement. This ability of deployment automation by the GitOps tool makes for an easy user adoption process as well.

The main integration point that makes an automated GitOps process beneficial is the git repository. By treating the git repository as the source of truth for the desired state, the GitOps tool enables easy synchronization between the user and the cluster through automated events.

However, any process or tool change needs to have an evaluation process that is relative to the level of impact the change will have on the users, teams, or business. In the case of originalist GitOps, the deployment automation process and tool change would significantly impact the teams that are administrating the tool, as well as the production environment that is being deployed to.

When looking at a GitOps tool such as Argo CD, which has some major benefits for any and all Kubernetes practitioners, the fundamental question to answer is "what is the tool actually accomplishing?" The automation of the deployment, the automated synchronization of the git repository and the cluster, and even the ease of install and initial set up are all great benefits. And yet, the manifest sprawl requirement, the lack of cross-cluster deployment capabilities, and the limited functionality across the entire delivery process can be significant enough reasons to look for an alternative tool or process.

Common originalist GitOps tools

The DevOps team felt as though they were back to square one with their research and implementation. Argo CD would've been the perfect tool for them to use if they had the rest of the delivery process figured out and automated. What they needed to do now was evaluate other GitOps tools in the market to see if they could solve the delivery needs better than Argo CD was able to. After an extensive search for alternatives to Argo CD, they discovered many different tools that all marketed themselves as GitOps tools or Argo CD alternatives.

The first on the list was Flux, which came from a company called Weaveworks. Weaveworks was actually the company that coined the term GitOps a few years ago and they built a tool that is similar to what Argo CD does. And, considering the developers are leveraging `helm` *charts, WeaveWorks has also built out a* `helm` *operator to work with their Flux GitOps tool as well. However, the manifest requirements and the lack of support for the rest of the delivery requirements are drawbacks that Flux and Argo CD have in common.*

Another tool that the team looked at was GitKube, which is similar in the sense that it can automate deployments to Kubernetes based on a change to a git repository, but the tool is advertised for testing changes to a manifest in a development environment, rather than for production clusters.

Another GitOps tool that seems to offer more features and functionality is JenkinsX. The DevOps team was already very familiar with Jenkins for CI and found it difficult to not associate the difficulties to JenkinsX But the documentation for JenkinsX seemed to show that the requirements were lighter than Jenkins's.

The company wanted to leverage a different cluster for each environment and was considering a multi-cloud strategy at some point in the near future, which were architecture designs that JenkinsX seems to support. The team decided that they would begin a GitOps proof-of-concept process with JenkinsX to see if it could fulfill all of their requirements.

Finding the right tool for an immediate need is simple, especially in engineering teams. If a tasks takes too long, requires a manual interaction, or causes consistent problems, an engineer will look to automate it. This process of looking for an immediate solution to an immediate issue can be considered a falvor-of-the-month-type approach to problem solving. Rather than solving the cause of an issue, the flavor-of-the-month solution looks to solve the immediate pain. This decision making process always leads to technical debt down the road since it rearely solves the actual problem.

Most DevOps teams have experience with a tool that was implemented to solve a small or inconsequential issue. Every veteran engineer knows what it is like to join a new team and immediately be hamstrung by all of the rushed decisions that were made by previous team members. In fact, most engineers assume and expect that any new team or company they join will have some level of limitation or process debt that is left over from previously rushed decisions.

Argo CD has become a very popular originalist GitOps tool, in that it automates the synchronization of a git repository and a cluster. Weaveworks Flux is a similar tool that attempts to accomplish the same requirement. In fact, Weaveworks and Argo attempted at one point to build a hybrid solution called Argo Flux.

Considering what Argo CD and Flux are able to do and where they sit, both are great options for automating Kubernetes deployments.

If a company is leveraging multiple clusters, multiple clouds, or multiple regions, they might need to look into a tool that has explicit support for those architecture requirements. Although tools like Argo CD and Flux could be used across multiple clusters or regions, the functionality of the tools would be too disparate to be beneficial. A tool like JenkinsX has support for multiple clusters and clouds built in and is also built to solve originalist GitOps requirements.

Summary

Originalist GitOps is defined as a tool that automatically deploys Kubernetes resources to a Kubernetes cluster by synchronizing the cluster with a git repository. The analogy in this chapter showed what originalist GitOps tools are able to do, some requirements of those tools, benefits and drawbacks, and the different types of originalist GitOps tools used in the industry.

The next chapter will explore the purist GitOps school of thought, showing how it is different than originalist GitOps and what tools can be used to accomplish a purist GitOps practice.

5
The Purist GitOps – Continuous Deployment Everywhere

The second school of thought can be categorized as **purist**. By adhering to the basic tenants of what makes GitOps beneficial, the purist mindset is broader in scope than the originalist school is.

As we mentioned in an earlier chapter, storing configuration code in a central source code manager is nothing new. But what makes GitOps, and especially purist GitOps, different from traditional practices is using source code management systems as a source of truth.

In this chapter, we're going to cover the following main topics:

- Purist GitOps basics

- Servers, containers, and serverless deployment as code

- Declarative files overload

- Benefits and drawbacks of purist GitOps

- Common purist GitOps tools

Let's get started!

Purist GitOps basics

The DevOps team had begun their trial run of Jenkins X. Even though the name is shared with the legacy tool that they attempted to use before, Jenkins X is significantly different. One main difference is that Jenkins X was built from the ground up as a Kubernetes-only deployment solution. Jenkins X has the ability to even create and configure brand new Kubernetes clusters, which could be extremely beneficial when you need to scale out to other regions.

Connecting Jenkins X to the GitHub repository was an easy task, and the first deployment was executed almost immediately. Although the tool still had a problem with the fact that nothing in the manifest was changing, it was able to deploy across multiple clusters when a manual trigger was kicked off. Of the issues that the team encountered with Argo CD, the ability to easily deploy the same application across multiple clusters was very important. The development teams would have to be instructed to use a better tagging system to solve the manifest change issue. This change would not only be pushing an artifact with a designated tagging schema, but also requiring a manifest update for each new artifact.

But just when the proof of concept seemed to be moving along well, the engineering leadership made some scope-altering decisions. The first of the decisions was that they would need to keep some of the application servers around. This was because of the required support for their on-premises-based customers who are not able to leverage Kubernetes internally. The other major decision was to leverage serverless technology where possible, to lower costs and increase application speed. For the DevOps team, this meant that they now had to support applications across legacy, containerized, and serverless architectures. In addition, they would need to decide if it was better to leverage multiple tools for the delivery and deployment process, or just one.

One major benefit of this new support scope was that the DevOps team will now have more time to come up with a solution for the entirety of the delivery and deployment phase. The needs of the deployment automation process still remained the same, which meant that they could still leverage a GitOps style of configuration and deployment. But, because the architecture had now expanded beyond Kubernetes, Jenkins X would no longer be a suitable solution. The support teams in the company had different tools for log monitoring, metric monitoring, ticketing, testing, and so on. Building, maintaining, and administrating multiple tools to execute the same task was not something the team wanted to add to their regular tasks. Instead, the team decided to analyze what they liked about the GitOps process and see what tools could handle GitOps across the different architectures.

The general premise of GitOps is that all of the configuration for the desired application state would live in a source code manager. The advantage of tools such as Argo CD and Jenkins X was that they had native execution capabilities out-of-the-box. But the ability to execute a deployment was not very difficult, since GitHub already had the ability to execute an event when a repository event was triggered. Essentially, for the team to achieve GitOps, they had to enforce the purest sense of GitOps for all applications. The core requirements for the desired tool would native integration into GitHub, the deployment across multiple architecture types, Git event triggers, and leveraging desired state files.

The architecture design of an application can be difficult to define. All applications, even mature ones, will evolve over time. In fact, a common trend in the industry around continuous processes is to have continuous iterations of the application. Some of these iterations might be breaking out code into containers, while some could be reducing functions to serverless capabilities. Not every company in the industry has the ability to flip a switch and convert their entire application into a single architecture type. Therefore, the ability to provide scalability and reliability is linked to repeatable automation. And the only way to achieve repeatable automation is to acquire standardization in the execution and configuration.

When approaching these ever-changing application architectures, it is important to note that a company should not look for a single tool that does everything. However, a single tool that does everything is not always bad. The balance between number of tools and required functionality is important understand. Proofs of concept that are well-scoped with success criteria will help with finding the balance between over-extending a tool and over-extending the team. In many cases, teams will over-extend tools before they would move to a tool that meets all of their needs. In the case of application architecture support for deployments, one tool is better than multiple, since the scope of deployments is relatively small. However, the common trend in the industry is to leverage a tool such as Jenkins for continuous integration and then over-extend it by adding significant scripting to make it support every possible use case. Jumping the chasm between continuous integration and continuous deployment should be a multi-tool project. And, in the case of Kubernetes, leveraging Jenkins for continuous integration and Jenkins X for continuous deployments is better than over-extending. If a DevOps team wants the benefits of GitOps but has to extend beyond the Kubernetes architecture, they can leverage a purist implementation of GitOps.

Originalist GitOps is built to synchronize a cluster resource with a manifest is Git. The purist implementation of GitOps includes Kubernetes in its scope, but it attempts to include every other architecture type as well. Similar to the Kubernetes operator pattern of originalist GitOps, purist GitOps requires a desired application state to live in a Git repository. But, contrary to originalist GitOps, the desired application state is not enforced upon the current application state through synchronization. Rather, the Git repository is a way to store the desired configuration state as a point of reference and the execution engine becomes the source of truth. When the execution engine sees a difference in the desired application state, typically through a Git event, it will attempt to deploy the desired state into the environment. One of the major advantages of this style of GitOps is that a failure does not cause configuration drift.

Configuration drift is when two or more items that should have the same configuration are not synchronized. For example, if there is a failure in the cluster, a failure strategy might have to be executed on the cluster directly. This means that the configuration in the Git repository has drifted behind the configuration in the cluster. This failure to deploy can be a difficult issue to solve when the failure is in production or on a high-volume application and requires a quick solution. However, if a GitOps tool operates as the source of truth, rather than the Git repository, configuration drift is less of a concern.

Purist GitOps, in concept, has been practiced, in some way, ever since Puppet, Ansible, or Infrastructure as Code tools became popular. Essentially, the ability to store a configuration file in a source code repository and leverage it for a deployment is what makes purist GitOps enticing. However, because purist GitOps has a larger scope of support, the criteria for choosing a tool is much different than originalist. Instead of basing the decision on supported architecture, a purist GitOps tool should be chosen based on what the tool was built for. If the application is legacy in nature and will never move off of legacy architecture, then a tool that was built for legacy deployments will suffice. However, if the application architecture is continually evolving and might leverage servers, containers, or serverless technologies, then a tool that is under constant improvement by a large set of contributors should be considered. This becomes especially important to consider when a wide array of technology types are being used, which most organizations today have.

Servers, containers, and serverless deployment-as-code

The main pipeline that currently being used, at least in part, was in Jenkins. Although Jenkins is not considered a GitOps tool, the ability to leverage the core GitOps practices through it is possible. Through the use of Jenkins Pipelines, the team could design the execution engine in Jenkins and allow users to maintain a single declarative configuration file, known as a Jenkinsfile. This Jenkinsfile would live in each Git repository, meaning that the integration process and deployment process configurations could be defined and stored per team. And, since the DevOps team was already familiar with Jenkins, they wanted to evaluate Jenkins as a potential purist GitOps tool.

Initially, the GitOps setup in Jenkins required the ability to build any type of artifact. Natively, Jenkins can build the artifacts for their legacy environment fairly well. But, for containerized artifacts, the team would have to either build that process or enable a plugin for it. Although there were security and maintenance issues with leveraging an open sourced plugin, it was the quickest path to the desired outcome. Next, the GitOps process would need the ability to deploy the artifact to the lowest pre-production environment based on a Git repository trigger. The Jenkins build that executed the deployment would not be too difficult to add, but the different artifact types required individual pipelines to be built. This one requirement added significant complexity to the deployment process.

In order to build out the required builds and pipelines to operate a GitOps process for each artifact type, the DevOps team needed to get a non-GitOps process working first. The initial step for them was the deployment process for containers. This would be the easiest piece to start with since they would only need to tweak their current pipeline. The goal of the pipeline is to allow a new container to be deployed to Kubernetes based on a Git event. Therefore, there were only two things that the DevOps team had to implement in their current pipeline, with multi-environment support at the trigger.

To support a multi-environment deployment for Kubernetes, the team found that leveraging a service account token and the Kubernetes master node URL would be the easiest option for repeatability. This allowed the team to use the URL and token combination for each cluster at execution time on the same build.

The next step was to construct a pipeline that would run the same Jenkins job in sequence, with a health check at the end of each step. Once they scripted and tested the build against one cluster, they needed to make sure that they could execute the same build against the other clusters. After the pipeline had been constructed with the reusable build, the team then needed to work on how they would get the trigger to run successfully. The hope was that a new Git event would trigger the deployment, but they needed to figure out a way to ensure that the new container was built, and that the Helm Chart was updated correctly. But, as they were looking at the trigger requirements, they encountered a few concerns.

Jenkins was their current integration tool, which generated their containers and Helm Charts. The DevOps team wasn't comfortable with over-extending Jenkins into deployments. They could still upload the Helm Chart file to the Git repository, but they wouldn't have to wait on the Git event to trigger the deployment. Another issue was creating a vast number of triggers to handle every container artifact and chart for each microservice. They wanted to make their process as easy to maintain as possible in order to achieve scalability and reliability across all deployments.

The solution that the team came up with was for the developers to maintain their own Jenkinsfile in their Git repository. The Jenkinsfile would include the specifications for the build process, environment configurations, and testing requirements. The update of the Jenkinsfile would trigger a full integration and deployment process in Jenkins. The integration process would produce a new container artifact version and Helm Chart and upload them to the desired endpoints. As soon as the upload was finished, Jenkins would then execute the deployment against the different clusters in the predefined order. Each cluster that the Helm Chart was deployed against would require a manual check by the developers before they could move on to the next environment.

Once the DevOps team had the deployment flow figured out for the container-based application, they decided to move on to the serverless-based option. They wanted to keep a GitOps process approach for the serverless application, which would be much simpler since it is a cloud-native offering. The main requirements would be the serverless artifact, which was a ZIP file, the artifact configuration requirements, and any ancillary pieces to the deployment. Having Jenkins compile the code into a ZIP file for the artifact was very simple, especially since they could upload the ZIP file to a storage service in the cloud that would host their serverless application. The main issue that the team faced was how they would define the application configuration and ancillary requirements. Their search for a solution resulted in the need to leverage an Infrastructure as Code solution, such as Terraform or Pulumi. Since the cloud infrastructure team was already leveraging Terraform, the DevOps team wanted to standardize on that.

Similar to the Jenkinsfile requirements for the container-based application, the developers would maintain a Jenkinsfile for the serverless application as well. The main difference would be that the Jenkinsfile for serverless applications would house the configurations that would be passed into Terraform, rather than a Helm Chart. And, to keep the GitOps style for the serverless deployments, the Terraform files would be generated and stored in a Git repository next to the application code and Jenkinsfile.

Since serverless applications don't require as many configuration options as a container-based application does, creating the Terraform files would be much simpler to do. But now that the team had a clearer line of sight to the desired end state for containers and serverless applications, they would need to turn their attention to the server-based applications.

The original legacy application was still being supported and deployed quarterly. But the engineering leadership team wanted to convert the legacy application into something that could be deployed more consistently. For this, the team had to figure out how to best deploy the application across every server with little or no downtime. Also, the team needed to figure out how to structure the servers so that they were consistent, allowing for the deployment process to be declaratively defined. Lastly, the team had to figure out how to leverage a Jenkinsfile to pass the required configurations to the deployment process without overloading the developers.

The DevOps team started to lay out their different plans for the serverless applications, the container applications, and the server applications. As they mapped the relationship between the files and pipelines, they saw a labrynth of code. They now understand what it means to have spaghetti code.

Any engineer that has been working in the DevOps space has some knowledge of Jenkins. Many companies in the industry are still leveraging Jenkins as their integration tool of choice. One of the reasons for the popularity of Jenkins is that it offers significant flexibility and customization. Whether the Jenkins user builds their own plugins or leverages community plugins, the ability to extend Jenkins to meet any need is extensive. However, one thing that plagues the DevOps community when it comes to using Jenkins is customizing Jenkins to execute a process that it was not designed for. This over-extension of Jenkins has become commonplace for most users in the DevOps world as the implementation of containerized or serverless applications continues to grow.

As companies slowly migrate from their monolithic or legacy applications to a cloud-native architecture, the need to enact a standardized process for integration, delivery, and deployment becomes a central issue. The recent flood of Jenkins plugins and tutorials around cloud-native support is proof that this standardization requirement is the new norm. Those who have had to support an entirely monolithic or server-based application are able to easily accomodate the quarterly deployment process. And those who have taken part in the move from monoliths to microservices are able to resonate with the issue of multi-architecture support headaches.

Server-based applications have always allowed for some level of consistency, since the artifact would need the same configuration on each server. The main issue is with how traffic gets to the application on the server, especially when a new deployment is required. Being able to check the health of the application on the servers across multiple regions has never been an easy process. Standardizing the deployment and testing the new artifact is paramount to having a repeatable, reliable, and scalable process. A common practice to support this standardization is to treat the server as an immutable architecture. What this means for the developer is that the main server, such as a virtual machine, and the application server, which might be Tomcat or JBOSS, are set up in the default manner and do not change. Therefore, the main agent of change is the new version of the core application artifact, such as a JAR or WAR file. The process of deployment is relatively simple as a result of this immutable architecture. The deployment process would simply need to stop the application server, change the required configuration files, copy and set up the artifact in the correct folder, and then restart the application server. This deployment style makes for an extremely scalable and repeatable process since the folder structure is the same on every machine. Additionally, the immutable architecture would not require any network or hardware changes as the deployment process is executed. The only part of the process to then figure out is how the server selection process will be automated. This deployment repeatability would be rather easy for Jenkins to execute, especially since Jenkins was built for monolithic and legacy applications.

Containerized applications in Kubernetes are, by design, significantly easier to automate and scale. The underlying controllers in the orchestration engine allow for the user to only declare the end state of the application. But, as was mentioned previously, Jenkins was not originally built for containerized applications. The ability for Jenkins to execute an integration build for a container artifact requires a plugin and some scripting. Over-extending Jenkins into the containerized world becomes even more difficult when considering the Kubernetes manifests. Because Jenkins was built for server-based applications, it is very easy to get a configuration file or run a configuration script on a server. But the interaction between Jenkins and Kubernetes is significantly different. Since Kubernetes is mainly updated through a combination of resource manifests and `kubectl` commands, Jenkins would need a way to get the manifest, access the cluster, and execute the required commands on the cluster. However, the generation or mutation of the resource manifests is where Jenkins shows significant gaps. Because Jenkins does not have a native `build manifest` option an administrator or developer would need to create a manifest template for Jenkins to manipulate. In some cases, Jenkins is configured to use a very basic template and leverages a long list of flags for the execution command. In other cases, developers will pass in deployment requirements via a Jenkinsfile in their Git repository, which Jenkins will use to output a values file. If the DevOps team wants to leverage a GitOps style process with this Jenkins style, they need to decide the best way for Jenkins to generate the manifest. And, once that manifest is generated, Jenkins would need to upload the container artifact and the resource manifest to the correct endpoints before the deployment could begin.

The process of supporting serverless applications is somewhat similar to containerized applications. The artifact is relatively small, and their configuration and deployment are typically declarative in nature. The main difference between the two is that containerized applications leverage multiple resource manifests and serverless applications only need one or two. This difference is the result of containerized applications still interacting, at least at some level, with the underlying infrastructure. However, with serverless applications, the artifact consists of a small function that does not require infrastructure configurations. The configuration requirements for a serverless application are centered more around the resource requirements for the function to run, traffic access requirements, and any trigger information. Since most of the configuration for serverless applications is more focused on how the cloud serves up the application, it is common for the configurations to be applied through an Infrastructure as Code process. But, as was shown with regard to Jenkins and containerized applications, Jenkins would need the ability to create the required manifest and artifact. The DevOps team administrating serverless support through Jenkins will have to decide how to execute the required deployment and file creation commands. Although executing and creating the files for serverless via Jenkins is the same requirement for the Jenkins flow for containers, the outcome is significantly different. The other issue with leveraging Jenkins for container-based and serverless applications is its abstraction layering.

Abstraction can be thought of as how many times a command must be translated before the desired action is completed. In the case of Kubernetes, the first layer of abstraction is executing a `kubectl` command. The next layer of abstraction is the command being passed from `kube-api-server` to `etcd` for `controllers` to see. Another layer down is the process of `controllers` passing the desired commands to `kubelet` and `kube-scheduler`. The last layer of abstraction is `kubelet` implementing the desired state on the *node*. By interacting with Kubernetes, the user is removed from the final execution by at least four layers of abstraction.

Jenkins typically operates either as the only abstraction layer or interacts with a single abstraction layer. When deploying to Kubernetes, Jenkins has to operate through multiple layers of abstraction, which can result in issues that Jenkins does not know about or doesn't know how to handle. And when you're pairing serverless applications and Infrastructure as Code solutions, Jenkins becomes even further removed.

But assuming that the team understands these issues and limitations and still wants to leverage Jenkins, the next major issue to tackle is the most common problem with declarative language processes. How does a team make sense of the sudden overload of declarative files?

Declarative files overload

Jenkins, Terraform, Kubernetes, Tomcat, servers, serverless, containers, pipelines, builds, and plugins. These are the tools, solutions, and architecture types that the DevOps team had to implement in a scalable, repeatable, and reliable way through GitOps. Implementing GitOps with just Kubernetes had its own issues but was significantly easier than trying to leverage a GitOps process for servers, serverless, and container architectures.

Initially, the team thought that by having the developers maintain a Jenkinsfile for every service repository, they would reduce the administration requirement for deployments. This was a rather easy setup when considering a traditional server-based infrastructure. The developers would simply add the desired configuration information that the application would need to their Jenkinsfile. Jenkins would then be able to pass the configuration requirements to the server at runtime, execute the restart and copy commands, and be finished with the deployment. The DevOps team would only have to maintain the files associated with the Jenkins pipeline and have a Jenkinsfile template for the developers to maintain. The DevOps team would have to maintain a set of 10 or so files for the core pipeline and one file for every repository.

With the support scope that includes Kubernetes and containers, the team had to build a different pipeline for their execution. Leveraging a Jenkinsfile was still the preferred method for gathering any configuration specifics, which the developers could then maintain. But, adding to the Jenkinsfile requirements, the Jenkins pipeline would require an access file for every cluster and environment to deploy to. Additionally, the Jenkins pipeline would need the ability to generate a values file for the Helm Chart to use. By leveraging a core Helm Chart for all microservices to use, the team could limit the Helm Chart sprawl. But since the required Kubernetes resources varied across each team and microservice, the DevOps team had to figure out whether to have one core chart or multiple base charts. If they chose one core chart, then every chart would have the same type of resources, even though they may not need them. If they chose to use multiple base charts, then they had to figure out a way to select the desired chart at runtime. The most secure option was to leverage multiple base charts, since that would provision a minimal set of resources per microservice. As a result of these container-based deployment requirements, the DevOps team would to support and maintain at least 50 core files.

After adding serverless application support, the DevOps team had to build another deployment pipeline. And, similar to the containerized applications needing Kubernetes manifests, the serverless applications would require Terraform files for each service. The DevOps team could build out and support a core set of Terraform files and allow a Terraform values file to dictate the different configuration requirements. By leveraging a Jenkinsfile for the serverless service repositories, the Jenkins pipeline would be able to generate the required values file at runtime. Supporting the serverless applications in Jenkins would add an additional 30 or more configuration files to the mix.

The team would also need to supply and maintain the appropriate credential files for the servers, the clusters, the cloud, and for Terraform, resulting in an additional 8 to 10 files. Lastly, the DevOps team would need to consider how they would maintain any Jenkins plugins or shared template libraries.

By the end of the pipeline design process, the DevOps team figured that they would need to create and maintain at least 100 core files. But when they considered all the Terraform values files, the Helm Chart values files, and Jenkinsfiles, they figured that they would need to support between 300 and 500 declarative files in total. And to make matters worse, each file referenced at least one other file, with some referencing up to 15 different files. This meant that if there was an issue with a deployment, the troubleshooting process would be like trying to unravel an entire plate of spaghetti, without breaking a noodle.

Declarative processes have been revolutionary for the world of DevOps. Something as simple as declaring the desired outcome and allowing an execution engine to deliver the outcome for you. Prior to the declarative process becoming popular, the majority of engineers and administrators relied on imperative processes. Building out the entire execution and configuration in one main file resulted in non-scalable and brittle processes. But any team wanting to leverage a declarative process must figure out how they will handle the potential declarative file requirements.

If an engineer is only deploying one artifact and with minimal configuration requirements, a small and simple script can handle everything they need. The moment that other artifacts are included, the administrator must decide whether to increase the size of the file or increase the number of files they must maintain. No matter which tool or solution a team decides to leverage, the biggest concern is related to whether file size or file count is the preferred support method.

Whenever a team leverages multiple tools or solutions, they must immediately support multiple files. Typically, this is a result of the support for different languages and file structures that each tool or solution requires. Once a team encounters the multi-file requirement from the different solutions, they will then need to figure out how they will handle the file structure for each solution. A good example of this issue would be the structure of Kubernetes manifests.

Some Kubernetes administrators prefer to keep relative Kubernetes resources, such as multiple configuration maps, or `ConfigMaps`, in a single file. And other resources, such as Deployments, will have their own files. If a set of manifests requires a Deployment, a Service, an Ingress, a `ConfigMap`, and a Secret, the typical manifest file count is five. However, because of how a manifest file can be structured, the required file count ranges between one file, containing all resource definitions, and one file for each resource. This balance is not easy to calibrate and can often lead to massive file sprawl, especially as requirements and solutions grow.

When considering the set of tools and architectures that a DevOps team might have to interact with, the potential declarative file sprawl can seem almost unwieldy. Oftentimes, the desire to reduce a large number of small files that cross-reference each other can lead to a small number of large files that don't require much cross-referencing.

Another thing to consider when trying to find the balance between file size and file count is scalability. If the decision is to leverage a small number of large files, then scalability becomes more difficult, as the small number of large files can quickly become a large number of large files. Inversely, if the decision is to leverage a large number of small files, the large number will not increase as rapidly, but the cross-referencing between each file will require a constantly updated knowledge base of file relationships.

Benefits and drawbacks of purist GitOps

GitOps seemed significantly easier when the DevOps team only had to consider support for multiple Kubernetes clusters. There were multiple tools available to support the requirements in a native way. Developers wouldn't have to be trained on how they would configure any additional files. Setting up the tools had a minimal requirement on the teams. The only issue that the DevOps team had to consider was which GitOps tool was going to fit their scaling needs the best.

The increased support scope from engineering leadership meant that the DevOps team couldn't use the original set of GitOps tools, such as Argo CD and Jenkins X. And not only that, but the available tools that could support a GitOps style across all of the required architectures were non-existent. No tool had been built that natively supports a purist GitOps process for server-based, container-based, and serverless architecture deployments. The DevOps team had to resort back to Jenkins, which they had hoped they could move on from, considering all of the infrastructure and administration support Jenkins required. However, it seemed easier to take a tool with native legacy application support and extend it to support cloud-native technology, rather than build something from the ground up.

Adding to the headache of trying to extend Jenkins to support all of the architecture types, the DevOps team found themselves needing to add outside tools, such as Terraform. Although leveraging Terraform reduced the need to leverage other tools, it added significant file requirements. Adding Kubernetes and Helm Charts to the support mix only made things worse for the team. The thought of maintaining or supporting hundreds of different files, across different tools, with confusing cross-references was enough to lose sleep over.

But when the DevOps team considered the alternatives, the purist GitOps approach was definitely preferable. Being able to leverage declarative files for the executions meant that they could achieve the scalability and reliability that would make their lives much easier and less stressful. Having a single execution engine run all of the deployments allowed for easier deployment transparency across the teams and platforms. Each development team would only have a single file that they would have to maintain, making for less administrative efforts for the DevOps team.

Even though the number of required declarative files was daunting, the confidence that a purist GitOps process gave the DevOps team was encouraging. Once they had finished building the process out, they would then have a truly repeatable, reliable, and scalable deployment. But, before they began to build out everything on Jenkins, they wanted to do some quick research to see if any other tools existed that would allow for a more native approach to the purist GitOps method.

DevOps teams that leverage purist GitOps styles are able to benefit from the inherent nature of leveraging declarative files. Any tool or platform that has a **command-line interface (CLI)** or API allows for purist GitOps to integrate with it. Defining what the desired outcome of the execution is, without needing to design the actual execution process itself, makes for easy repeatability. And since purist GitOps is more focused on using declarative files stored in a source code management solution, any architecture can be maintained in a GitOps style.

With repeatability as an out-of-the-box benefit of purist GitOps, practitioners only need to make sure that the execution engine and deployment endpoints are immutable. If the execution process is immutable, then reliability across every execution can be guaranteed, since every execution will follow the same process. If the endpoints of the deployment are immutable, then the reliability of the application's end state can be guaranteed, since every endpoint will look the same. And if both the execution engine and the endpoints are immutable, then the reliability of its execution and end state can be guaranteed. Additionally, if the endpoints are created and configured in the same manner, then the scalability of the deployment process can be guaranteed as well.

The main drawback of a process such as purist GitOps is in the size and number of the required declarative files. If the declarative files are too large, then the configurability of each file becomes a usability issue. Alternatively, if the files remain small but the number of files grows, then tracking those files and file relationships becomes an administration issue.

One last piece to consider, since there are no native purist GitOps tools, is creating, administering, tracking, maintaining, and troubleshooting the files and file relationships. Each file must be created and tested by the team; only then can it be added to a repeatable process.

Even though there are no native purist GitOps tools available, there are a few other tools that might offer a better solution for those teams needing to support traditional and cloud-native deployment processes.

Common purist GitOps tools

Jenkins seemed like an obvious choice to fall back on for the DevOps team. They all had experience with the tool, and it was being used for the other deployment process that they are looking to replace. However, with the support requirements that the team now faced, as well as the desire to adopt a GitOps style of deployments, Jenkins was not the ideal tool for them. The team decided to research other tools that were able to support both cloud-native architectures and more traditional models.

Of the tools that they researched, they found that there were two main categories to choose from. Either they could look for a more modern integration tool, such as Drone or CircleCI, or they could leverage a multi-purpose tool, such as Puppet or Ansible. As they read the documentation around each of the tools available, what they found was that the integration tools were good for integration but would result in a similar setup as Jenkins did. Integration tools were good for that purpose, and over-extending the integration tools would result in a script runners. Alternatively, tools such as Puppet and Ansible were designed to be declaratively defined and were, at their core, execution engines. The more popular of the two options, especially considering it is actively in development by Red Hat, is Ansible.

Ansible has a built-in setup for declarative files called playbooks that a user can define an execution and file inheritance process in. The tool allows you to template the different files, allow for the files to be reused across different execution pipelines, and give to the entire spectrum of architecture support that the DevOps team would need significant flexibility. Additionally, Ansible and Terraform are able to work together for serverless application requirements. Lastly, because Ansible has a YAML format style for the playbooks, the learning curve would be easier since Kubernetes also leverages YAML files.

The main drawback of leveraging Ansible for a purist GitOps style of deployment is that the team would have to build most things from the ground up. With Jenkins, there were significant plugins that can be leveraged. But, considering how much easier Ansible would make cloud-native support, the team was leaning more toward leveraging Ansible over Jenkins.

Support for future automation requirements across the delivery pipeline was the only area that the team did not have confidence in. Jenkins and Ansible didn't necessarily offer built-in integrations with the majority of tools or solutions that they needed for their delivery requirements. Instead, the team had to either build the integration themselves, leverage open source modules or plugins, or keep the rest of the delivery pipeline a manual process. Considering the timeline, the requirements, and a desired future state of complete automation, the team looked for a way to implement GitOps not only on the deployment, but also for the entire delivery process.

Purist GitOps is significantly broader than the originalist GitOps model in supported architectures. But the main drawback of purist GitOps is the lack of native tooling. Although the theory makes sense, the need to build everything before purist GitOps can be implemented is difficult.

The most common form of leveraging the core essentials of GitOps, without implementing originalist GitOps practices, is in Infrastructure as Code. Tools such as Terraform and Pulumi are natively built and defined declaratively. Infrastructure as Code practitioners are already storing their code changes natively in Git repositories and pulling the different files and modules at runtime. In fact, that is one of the major benefits of these types of tools.

Currently, the best way to leverage purist GitOps is to build an in-house solution, either greenfield or brownfield. Building the tool in a greenfield way means that no development has been done yet and that the entire execution engine is coded from the ground up. Building the tool in a brownfield way means that some work has already been done, whether that's leveraging an external or internal tool, and then building on top of it. Building a brownfield purist GitOps tool would be like building on top of Jenkins or Ansible.

The hardest piece of this decision-making process for any team is that the desired future state of delivery automation seems to be almost in reach. By considering how to adopt and implement a purist GitOps methodology, the evaluation team must also consider how long they have left to achieve their ultimate delivery requirements.

Summary

This chapter explored the world of purist GitOps. The DevOps team, in our analogy, learned that adhering to the core essentials of GitOps, especially with the new support requirements, would offer the same desired outcome. Additionally, this chapter showed that the desired end goal of GitOps is to make a process repeatable, reliable, and scalable.

The next chapter will explore a third GitOps practice known as verified GitOps. This practice is the idea of leveraging GitOps practices across the entirety of the delivery pipeline, both in terms of execution and integration.

6
Verified GitOps – Continuous Delivery Declaratively Defined

The last GitOps concept that we will cover is a relatively recent practice known as **verified GitOps**. The essential difference between verified GitOps and the other GitOps practices is the desire to leverage GitOps for the entirety of continuous delivery, rather than only for continuous deployment.

One operating requirement across all GitOps practices is the need for declarative language files to be stored in a source code manager and referenced at runtime. And, depending on which practice is being adopted, the number of required files is heavily impacted based on the desired manual intervention. This results in a balancing act between manual intervention steps and the number of declarative files.

In this chapter, we're going to cover the following main topics:

- Verified GitOps basics

- Test, governance, deploy, verify, and restore as code

- One file, many files, or somewhere in-between

- Benefits and drawbacks of verified GitOps

- Common verified GitOps tools

Verified GitOps basics

The past few months of developing and implementing GitOps tools and processes have been illuminating for the DevOps team. The changes in scope from the business and engineering leadership were somewhat discouraging. However, as the team had to restart their efforts when implementing GitOps, they realized that the new requirements helped with futureproofing. Each process or tool that the team had looked at seemed like a great solution for the requirements at that time. But as they tested the solutions, especially with each scope change, they quickly realized that the inevitable future requirements would not have been met. Requirements such as cross-platform support, cloud-native and traditional support, and a desire for repeatable, reliable, and scalable processes were difficult to achieve with the other tools.

These requirements lead them to Ansible, which seemed like it was going to offer the best option for leveraging GitOps across all platform types. One thing that made Ansible so helpful was its ability to leverage declarative files in a native way. This is because Ansible operates in more of a scaffolding design. The DevOps team would need to fill in the pipelines with integrations and self-defined steps. If the team needed to execute a process on a server, they would still have to build the process out themselves. Then, when the process was built out, they could bring that process into Ansible for it to execute.

A major benefit of Ansible is allowing imperative commands to become declarative through the use of templating with variables. The DevOps team could build out they commands they wished to run, add the variables, and then reference those command files wherever they needed them. This templating capability meant that the team only had to build a step once and could then reuse it wherever they needed to. Because of this capability, the team had to figure out how best to leverage this reusability without causing too much confusion. This templating capability meant that DevOps could prevent repetitive development, but it also meant that they would have to balance the functionality with the potential file count.

By representing each deployment or integration function in code, the team would be able to have an execution engine that considers each function as a building block. Each execution could be somewhat unique because of the dynamic nature of templating and user-provided variables. But this would require an execution process to be file-reference heavy, resulting in difficult troubleshooting processes. The alternative is to build fewer execution processes, allowing for a significant number of input variables with less file referencing. The main concerns were the troubleshooting steps and total number of files required to fulfill their needs.

Since the team was considering how they would structure the execution process and files, they found that they would also need to solve secrets usage and user access. Audit requirements had to be included in the process, especially around who has access to the different environments and who deployed to each environment. And, in an effort to futureproof their process, they wanted to see if they could extend the typical deployment-only limitation of GitOps to some of the delivery requirements as well. But adding testing, change management, approvals, validation, and the rest of the delivery process requirements would not be easy.

One area that the team had to consider especially if they wanted to leverage a GitOps process, would be leveraging declarative files for tool integrations. Each integration would require its own file, which includes the configuration and connection requirements. And for every execution that includes the desired tool, the team would have to use another declarative file to define the desired interaction with the tool.

By representing every integration, interaction, execution, and configuration in code, the team would be able to achieve a more verified GitOps practice that would ensure a repeatable, reliable, and scalable approach to the delivery.

The most basic requirements for automating any process are the execution engine and a set of configuration files. The execution engine is the core appliance that enables a process to be automated, and the configuration files define the execution engine's behavior.

Every execution engine is purpose-built to solve a problem, which is reflected in its functionality. The purpose and functionality of the execution engine will lead to a set of standards and best practices for optimal performance and reliability. But a common trend in the engineering world is that execution engines become overextended. Users will often develop workarounds or hacks that allow them to break with the best practices, causing significant unreliability and often resulting in catastrophic system failures. Some execution engines become so overburdened that support teams have to work late evening and weekends to fix or rebuild broken platforms.

An example of overextending would be leveraging an execution engine that is purpose-built for deploying monolithic applications to servers. A common capability of these execution engines is the ability to run scripts from their servers. If a user decides to leverage the scripting capability to run deployment commands against non-server endpoints, then the process becomes error-prone. And since the native functionality is overextended, the tool is not equipped for proper troubleshooting or remediation processes. Overextended processes also result in significant tribal-knowledge and scripting requirements. And eventually, because of the required support burden, team members will churn, causing the entire process to stall and burn out. By avoiding overextension, the reliability of the tool and process can be better guaranteed.

If the execution engine enforces reliability and repeatability in a process, then the configuration files allow for scalability across supported platforms and teams. The developers of the execution engine can build out a configuration process that is both recommended and required. The configuration files should allow users to declare what the desired outcome of the execution should be, and any supported customizations are limited to what the configuration file can supply.

As the process that the group wants to automate grows in terms of support, requirements, and scope, the execution engine and configuration files will also increase. The execution engine will increase in terms of its size, maintenance requirements, administration requirements, and functionalities. However, what often is not addressed in this support scaling is redesigning the execution engine so that it fits a broader or more accurate purpose. Rather, the execution engine's initial purpose is extended to meet the new platforms and use cases, assuming that the new requirements can be addressed in the same way as the old requirements. Additionally, the configuration files will also increase, not just in the number of files, but also in their size and utility.

For example, in regard to a basic server deployment process, a script might be run by the execution engine and is intended to stop a process, copy application files and artifacts to the server, set up a required folder hierarchy, and then start the process again. The configuration files that are supplied to the execution engine would declare what to deploy, where to deploy it, and so on. As the number of servers increases, the execution engine can remain the same, but the configuration file count might grow with one new file for each server. If the required platform support grows to data centers and the cloud, the execution engine must now scale to support the nuances between the different network, platform, and security requirements. And if the required platform support adds non-server-based endpoints to deploy to, the execution engine and configuration files must now scale again. However, because the execution engine was originally purpose-built for server-based support, it will either need to be refactored entirely, or it will have to treat non-server deployments the same as server deployments.

Another thing to understand about automation is that the complexity of the execution engine and configuration files are tightly coupled with the complexity of the process that needs to be automated. Regarding the example of the server deployment, the process of stopping, copying, moving, and starting a process on a computer can be only a few lines of code. The moment that the process expands to include testing the artifact that was deployed on the server, the execution engine and related configuration files become significantly more complex. As such, detailing the different behavior and execution requirements are essential to the success of the automation outcome.

The continuous deployment process can be as simple as deploying to a single server. But it can also be complex such as deploying to a set of Kubernetes clusters across multiple regions. But as we learned in a previous chapter, the deployment is only a small portion of the delivery process. If a team wanted to automate their delivery process while also adhering to GitOps practices, they would need to leverage **verified GitOps**. Verified GitOps is a GitOps practice that focuses on providing repeatable, reliable, and scalable continuous delivery through GitOps.

Test, governance, deploy, verify, and restore as code

When the DevOps team was documenting the different stages of their desired delivery pipeline design, they wanted to include significantly more than just the deployment process. There was a desire to include QA testing, change management, verification of the production environment, and failure remediation. But shortly after those original design meetings, the team experienced an accelerated timeline and scope change. If they wanted to revisit the idea of automating the delivery process, especially with verified GitOps, they would need internal support.

The team found that it was better to start with a specific platform, build out the delivery pipeline for that platform, and then expand to the other platform support afterward. And because of the native execution capabilities of Kubernetes, it would be less of a lift for the team to build out the delivery pipeline for the containerized applications.

Initially, the team would have to solve the issue of Kubernetes manifest sprawl. If they could turn the main set of resource requirements for each application into a core template set, this would reduce the manifest maintenance. To accomplish this outcome, the team needed to consolidate Helm Charts and let teams leverage override files. At runtime, the execution engine would be prompted to fetch and deploy the manifest files from Git. The main requirement that had to be built was the fetch mechanism of the execution engine, since everything else was natively built into Helm's templating engine.

To use a verified GitOps practice the solution would need a set of declarative files. These files would give information about the Chart to fetch and deploy, access permissions, and environment configurations. And since integration and configuration files are not tightly-coupled to a specific pipeline, they could be referenced and reused almost infinitely. Once that process is defined in code, the team would need to build out the pre-deployment and post-deployment requirements.

The pre-deployment requirements that the team needed to add were the initial change management process and some preliminary security processes. To accomplish the change management requirement, the delivery pipeline would need to create a ticket with the required governance information. Each execution would have to add information about the artifact, environment, testing outcome, execution trigger information and any approvals. As for the security requirements, every execution would have to run a security scan on the artifact and the infrastructure, with the results being added to the change management ticket.

In order to define these steps in code, the team would have to build out tany files related to integrations, configurations, and executions. Similar to the declarative files for the Git integration the required supporting files for the change management and security processes could be built once and reused.

The next part of the delivery pipeline would include any post-deployment steps, such as ticket updates, deployment validation, and so on. But since the change management process was already built out for the pre-deployment steps, the team didn't have to build out that requirement again.

For the other parts of the post-deployment requirements, the team would have to figure out the best way to use a healthcheck to test for success, and what to do if the deployment fails. The easiest of these steps would be executing the restore process because Helm has a rollback feature already. The only part that the team would have to design is how to reference the previous deployment and automate the execution of the desired Helm commands.

In the case of testing the success of the deployment, Helm and Kubernetes can provide verbose logging to watch for any errors and will give the deployment status of the release. If the deployment status is successful the pipeline will be completed. In the case of a failed deployment, the pipeline should include an automated rollback process. The pipeline could reuse the release name from the deployment in the Helm rollback command. Being notified of the deployment status will also be a requirement, especially if there is a failure. At a minimum, the notification message should include the status and a link to the execution. Ideally, the message would include some information from the logs and error. Health checks and deployment validation would be the last steps to build out for the post-deployment.

The health check would not be too difficult if the application had a frontend or API. The pipeline would need to run the command on the desired endpoint and be able to validate the response.

The most difficult part of the post-deployment requirements would be the ability to verify the functionality against actual usage. The DevOps team wanted to leverage the monitoring tools that the engineers were the most accustomed to. One way to verify the deployment would be for the verification tool to send an alert when an issue occured. Most of the engineers were used to getting these alerts, so it would not be confusing for them. The desired point of integration is being able to tell the monitoring tool which deployment went out and getting information back. But even if they got information back, it would be difficult to discern what was good and was not. They would need to look for tools that could offer this capability and be added into their platform.

At the end of the requirements gathering process for the verified GitOps delivery pipeline, the team had a good understanding of what needed to be executed. They now have the difficult task of building out each execution step and declarative file that Ansible could leverage for the verified GitOps process.

Most companies are heterogeneous in their technology stack, meaning that they will have a wide range of different technologies to use and support. To automate a delivery process, every engineering organization will have a cross-team, cross-platform, or cross-cloud support requirement. The initial pain associated with this type of support is being able to adequately integrate and implement the individual steps and stages across each discipline. And each engineering discipline is its own market for a tool of some kind.

At some point, the desire for a general orchestration engine that integrates with the individual execution engines becomes very enticing. This is especially true when the orchestration engine can abstract away the operational knowledge requirements of the underlying tools. An orchestration engine, which is a tool that orchestrates a set of tasks or processes, is always enticing whenever multiple tools are used. A benefit to using an orchestration engine is in regard to reporting and transparency. Another advantage is that the administration requirements of the underlying execution engines can be managed by a set of **subject matter experts**, or **SMEs**. For many DevOps teams, Ansible is the orchestration engine of choice. Not only can Ansible integrate with other tools as an orchestrator, it can also be the execution engine when needed.

One main characteristic of an orchestration engine is its ability to behave as an abstraction layer that separates users from operation requirements. For example, Kubernetes has a built-in orchestration layer that abstracts the underlying operating procedures away from the Kubernetes users. Although a Kubernetes administrator can alter the operating procedures, the vast majority of users interacting with a Kubernetes cluster do not have to concern themselves with how Kubernetes accomplishes a task.

Representing the interaction between the underlying execution engines and the orchestration engine in code is a good way of understanding the verified GitOps concept. However, contrary to how originalist GitOps looked to the Git repository as the source of truth, verified GitOps leverages the orchestration engine as the source of truth. The assumption is that the different execution engines and platforms should continue to operate within their scope of purpose. The orchestration engine takes the user's desired state and passes it to the underlying layers, reports execution output, and then moves along the rest of the delivery flow. A core principal of verified GitOps is that execution engines and underlying platforms should be the source of truth. The declarative files in Git operate only as the source of the desired state.

As teams adopt verified GitOps, their orchestration layer will grow to incorporate many different platforms and execution engines. With this growth, they will need to figure out how best to balance their declarative file requirements.

One file, many files, or somewhere in-between

The DevOps team was finishing the design work on a verified GitOps pipeline and began documenting the requirements for the different integrations. They figured that they would need a set of declarative files for every integration, which included access and execution requirements. Then, they would be able to reuse those files in any pipelines that needed them. Because every provider, platform, and tool required these different sets of filesthe team would need to figure out the best folder hierarchy to store the files.

The team would then need to build out a set of declarative files that defined the required variables for integration interactions. Each interaction definition file would have to reference the appropriate integration files for access and permission requirements.

Next, the team would need to build some environment-specific execution requirements to enforce security and compliance standards for every deployment. The DevOps team would need to have a set of policy enforcement files to validate the compliance standards.

With all of the integration interactions codified, the team would need to build out the pipelines and associated triggers. The architecture of a pipeline was a point of contention within the DevOps team. Some of the team wanted to have a pipeline be one large file, which would reduce the overall number of files. Others on the team were worried about readability and preferred a higher number of small files that referenced each other.

No matter how the team designed the pipelines, they would inevitably have a pipeline of pipelines. If a group of related tasks were paired together into their own files, then a pipeline would be made up of building blocks. But this would have a significant amount of file-referencing, which can be very difficult to follow in code. The team needed to figure out how best to structure their pipelines to be scalable, repeatable, and reliable.

The biggest issue with any type of declarative file structure is the amount of work that a team has to contribute to get the maximum effectiveness out of the file. This is most commonly seen in Kubernetes manifests. Every Kubernetes resource can be declared statically in one extremely long YAML file. As templatization became a standard for Kubernetes administrators, these resource files would often be broken up into smaller sets of files for reusability. However, when the files became smaller, the number of files increased exponentially. An increase in the number of files leads to files referencing other files. When any file structure has significant inter-file referencing, the required knowledge to administer and troubleshoot issues increases.

The world of GitOps, regardless of it being original, purist, or verified, falls prey to this balance of manifest size and number of manifests. The end goal should be that the practice does not require repetitively building steps. But in a desire to limit the number of repetitive files, a verified GitOps administrator will be tempted to increase the number of small and unique files that reference each other. This is a delicate balance to get right since the outcome of the file structure will directly contribute to the readability and usability of the system.

As a team looks to adopt verified GitOps, they will need to decide who will build, maintain, use, and improve upon the system over time. By leveraging the large number of small files approach, the future maintenance, usage, and improvement of the system will be restricted to a small set of advanced users. Alternatively, by pursuing a small number of large files approach, the maintenance, usage, and improvement of the system will be easier.

Benefits and drawbacks of verified GitOps

By working through the nuances of verified GitOps and how to support all of the potential platforms and tools that the delivery pipeline requires, the DevOps team has, yet again, found themselves in the position of questioning their tool choice.

Even with all of the testing and research that they conducted, the team found that the building and maintaining requirement was too heavy. Ansible is highly customizable, but requires its users to build out every step in every file. And since there is no auto-generation capability, the overall file building requirement was massive. If the company only needed to support one or two platforms, the workload would be bearable. But because they had to support many different platforms, tools, and use cases, the build requirements were daunting. They needed to make a quick decision as to whether to search for another potential solution or start building out the Ansible process immediately.

But a quick search for a verified GitOps solution gave them zero results. Although there are a significant number of tools that state themselves as being GitOps tools, there was no time to analyze every single tool in that list. Instead, the team figured that if a tool advertised itself as a GitOps tool, showed that it had support for the platforms and tools they needed, and had some form of pipeline code, they would consider it.

These requirements whittled down the list to a small set of tools, which made the analysis easier for the DevOps team. Then, what they needed to understand was how the tools supported the platforms and underlying tool integrations, and how was the pipeline code was leveraged. If they were going to recommend buying a solution, then it needed to have native and intuitive support for their requirements.

Tool documentation would give more insight into how different integrations or tasks were supported. Some of the tools advertised support for a platform or integration, but really meant that a user could build and run scripts. Since that was an issue that they had with their Ansible setup, they decided to move those solutions to the bottom of the list of potentials.

Diving deeper into the solutions that were left, they found a tool with a signficant number of native integrations, GitOps support, and declarative configuration files. The documentation seems to indicate that the tool auto-generates the configuration files and can push them to Git. One drawback with the tool was that the testing and security tools that their company was using did not have native support. However, the platforms were all natively supported, as were most of the verified GitOps requirements. The team would need to download the trial, configure the solution, and see if it would work for them.

Understanding the benefits and drawbacks of any practice is paramount to success. The intent should be to capitalize on the benefits and be fully aware of the drawbacks before making any decision about adopting the practice. And as has been shown throughout this book, the drawbacks of GitOps can range from potential non-starters to mere annoyances. For example, if a team needs to support any platform outside of Kubernetes, then originalist GitOps is a non-starter for them. However, with purist GitOps, there may not be a way to directly support the platform, but the user can work around that issue.

Verified GitOps falls prey to a common GitOps drawback, which is the need to support and build out some set of declarative files. The balance between a few large files and many small files is an annoyance that should be understood when approaching verified GitOps as a potential solution.

Another drawback is that there are no tools marketed toward verified GitOps or purist GitOps. But even though there are no tools purpose-built for these GitOps practices, this does not mean that the practices should be abandoned. There are some tools that exist that allow the execution engine to be extended to support verified GitOps, without it being overextended or overburdened.

Verified GitOps has some major benefits, when properly instrumented. If every integration point is defined declaratively, then the usage of that integration can be more reliable and repeatable. If a team wants to implement a change management process that requires significant data input, an automated approach is the best way to ensure uniformity. By declaratively defining the change management automation with the desired set of data inputs, the outcome will be a reliable process that is automatically repeated and can be scaled across every execution as needed.

The biggest benefit of verified GitOps is the confidence across the entire delivery process that comes from a repeatable, reliable, and scalable practice. Every security, compliance, testing, verification, and deployment requirement can be enforced without manual intervention by users. And as every engineering team knows, fewer manual requirements in any process immediately reduces mistakes, bugs, and bottlenecks.

Common verified GitOps tools

The process of designing the requirements for verified GitOps gave the team a better understanding of where they should spend their time. The DevOps team has experienced building an in-house solution before, mainly avoid the costs of buying a tool. But they have also experienced the significant administration and maintenance effort associated to building a tool. Although Ansible had offered a wide range of customization capabilities, it was essentially a tool that they would have to build out and maintain. The other problem with Ansible was that although the files could be stored in a Git repository and pulled at runtime, the tool required the Git repository to be pulled down before every execution.

After performing market research for a verified GitOps tool, the team found **Harness**, *which had a very promising solution. The tool allowed the team to support Kubernetes, serverless, and server-based platforms. It had a native code conversion process that turned the pipelines in the UI into code, which could be stored in Git. Lastly, the team found that the SaaS style of the tool allowed them to maintain significantly less hardware and software in their environment.*

Another massive benefit is that Harness allowed them to quickly adopt a verified GitOps approach for the delivery requirements. They could easily define their integrations, tie those integrations to deployments steps, and enforce all of the executions based on their security and compliance restrictions. These benefits were all significantly enhanced because the tool has the ability to leverage built-in variables in a host of different areas, similar to Ansible.

Even with all of that, there were a few drawbacks to the tool, one of which is the lack of native integrations with their testing and security tools. But the tool did offer the ability to run scripts to supplement that issue. The same problem exists with the security tools that were required, such as artifact scanning. Although the scripting piece would allow for pseudo-integration, the team would have to build it out themselves.

One last drawback for the team was the fact that the tool was not free. They knew that trying to get a purchase request through the engineering leadership and company was not going to be easy. Setting up the solution was simple, but finding a justifiable reason to purchase a tool over building out Ansible was a different story. What they needed to do was have one group from their team start to build out the Ansible requirements while the other group worked on a way to justify purchasing Harness. Regardless of the outcome, they needed something in order to meet the required deadline for support, whether that was deployment automation or delivery automation.

Verified GitOps is a practice with a broad scope of purpose. Because of that, any tooling that is available will need to support a broad scope of purpose. An example would be something such as Jenkins, where its original scope of purpose is integration, but could allow for more. Using Jenkins is not uncommon in the industry, and the same can be said for any integration or script running tool. But what needs to be considered is the engineering work required to achieve the desired scope of purpose.

An alternative to overextending a tool such as Jenkins would be using a generic execution engine, such as Ansible. Ansible is an open source tool that is commonly compared to Jenkins, Puppet, Terraform, and others. Considering the wide range of tools mentioned, it is clear that Ansible is much more than a simple provisioning or configuration management tool. With the use of YAML files, plugins, and a central execution engine, Ansible meets many of the requirements of verified GitOps, even though it is not marketed as a verified GitOps tool.

Some companies forgo the tool route entirely, choosing instead to build a solution in-house. In most cases, companies do not associate the time and effort to build a tool with a hard cost, even though building and maintaining a tool can be a heavy requirement. Scaling an in-house built tool can not only cause higher support costs but can even lead to attrition of employees. A good way to understand the build cost is to consider the time associated with building, maintaining, documenting, improving, administrating, and hosting the tool. If four engineers, paid $100,000 a year, spend 6 months building the tool, that is $200,000 that's been spent just on building the tool. Maintaining and administering the tool might take 2 to 3 days a week, or 1 to 2 weeks a month, resulting in an additional $23,000 to $47,000 a year. The documentation, training, and iterating on the tool will cost about the same as the maintenance, which is another $23,000 to $47,000 a year. Finally, the hosting costs of the tool, especially if it is hosted in the cloud, might run an additional $5,000 a year or more, depending on scaling needs. This means that the first year will cost between $230,000 and $250,000, with a run rate of $51,000 to $100,000 on the low end.

The alternative to an in-house built tool is to buy a tool, but that comes with its own issues. Any purchasing decision has to consider the hosting and maintenance requirements, the growth of the tool and relative licensing expansions, and break-glass scenarios. Harness is a tool with a scope of purpose that covers the delivery process. But, because it is a **Software as a Service (SaaS)** solution, it must be licensed. Although it has a lower hosting and administration cost, it has a higher hard cost, which are any costs that a company sees on a purchase order. Alternatively, soft costs are any costs that can be hidden in employee hours or hosting requirements.

Although Harness does not advertise itself as a verified GitOps tool, its scope of purpose covers the same concepts. The ability of Harness to define every integration, permission, execution, and use case in declarative files, and then store those files in a Git repository, makes for an easy setup and configuration process. The drawback of Harness is that it doesn't have a native integration with every possible solution or process on the market. Therefore, anyone looking into a tool such as Harness needs to understand the work required to build out support for missing integration points.

Every time there is a requirement for a new tool or process, it is of extreme importance to understand the soft and hard costs involved. The cost to build or adopt, the cost of maintenance, the cost of administration, the cost of scaling, and licensing costs are all things that need to be considered before making any decision.

Summary

Verified GitOps is a practice that leverages declarative language files to solve the automation around the delivery process. By codifying the integrations, permissions, and executions, the team that adopts verified GitOps can quickly ensure it is repeatable, reliable, and scalable for every delivery. But since there are no tools that are purpose-built for verified GitOps, the options are limited to either a delivery-based tool, such as Harness, or a general script-running tool, such as Ansible.

In the next chapter, we will cover the best practices for deployment, delivery, and GitOps. The adoption of best practices will not only result in meeting industry standards, but also allow for a team and company to avoid tool lock-in.

7
Best Practices for Delivery, Deployment, and GitOps

All the previous chapters were focused more on the hypothetical concepts surrounding GitOps, delivery, deployment, and the software delivery life cycle. The goal of this chapter is to provide a simple overview of best practices, why best practices are important to consider, and how to implement GitOps to fit those best practices.

It is rather uncommon to have an unbiased approach to best practices, especially in the world of engineering. Every person who has worked with an engineering team knows of a person, or even a group of people, that will not be persuaded away from their original idea. This chapter will hopefully enable some best practices you should consider before you consider solutions.

In this chapter, we're going to cover the following main topics:

- The purpose of best practices
- Continuous deployment considerations
- Continuous delivery considerations
- Where GitOps ties in

The purpose of best practices

When the DevOps team first started out on the journey of automation, they were trying to script deployments to application servers. As the server pools and customer base grew, so did the need for more reliable deployment methods. These requirements changed rapidly as the company moved toward a cloud-native approach. The desire to adopt container-based and serverless applications with cloud infrastructure became the main drive behind the move for increased automation efforts. As the DevOps team built out a potential deployment solution for containers and Kubernetes, the directives once again shifted to include a broad range of platforms and architectures. Adding to the consistent scope of support shifting was a deadline that remained relatively consistent. This meant that the DevOps team had little time to document and plan, but rather had to implement a solution that was an inexpensive, high-quality product, and did so in a short period of time. And as any engineer will attest to, in the triangle of low-cost, high-quality, and short time, you get to pick two.

As the team was trying to build a business case for purchasing a tool, they realized that what they ultimately needed was a process that any tool could fulfill. The DevOps team consistently found themselves scrambling to get something completed. Every time a new platform, architecture, or scope changed, the team rushed to get a basic understanding of it in order to add support as quickly as possible. The general premise they were operating under was to take an artifact and put it into the desired environment with any required configurations. But what they had not considered was that the process to deploy was more important to understand and define than the actual deployment activity.

By breaking down the desired deployment and delivery requirements, they could develop best practices that must be fulfilled, regardless of the tool they chose. If they had considered and understood these best practices at the beginning, then some of the time they spent on a proof of concept would not have been wasted. Although they were able to avoid testing out tools based on integrations or support capabilities, the team did not come to an understanding of those requirements until it was too late.

As this best practice consideration was being discussed among the team, they realized that building out true best practices was essential before they chose a tool such as Ansible or Harness. By defining their best practices, and not just integration capability requirements, they would be able to easily move away from any tool that they implemented. But by not defining best practices ahead of implementing a tool, the team would then become bound to a specific tool and its limitations. Being bound to a tool means that any attempt to migrate off the tool results in finding a tool similar or one that can be built out in similar ways. With this new-found understanding of the need for best practices, the team needed to consider what their best practices would be and how to implement them with a verified GitOps tool.

Two of the biggest concerns that an engineering team must confront when choosing a tool are **lock-in** and **administration**. Lock-in is often considered when working with a vendor; this is also known as vendor lock-in. This is the idea that if a company or team purchases a solution from a vendor, then that vendor will lock that group into their ecosystem of tools and make it difficult to leave. The benefit of support and ease of administration is usually the trade-off when attempting to avoid vendor lock-in and leverage a non-vendor solution. However, lock-in is a concern regardless of having a vendor, since any group can be locked into the functionality or nuances of any tool.

The concern of administration is a broad one. Administration can refer to the requirements of a core group of power users that install and configure the tool, the ongoing maintenance of the tool, and the work required to supplement the gaps in a tool.. These administration efforts must offset each other for the adoption of the tool to be effective. For example, if a team spends a significant amount of time on setup and configuration, then the tool should be easy to use and learn.

In many cases, these two concerns are the driving factors for why a tool or solution is chosen and implemented. Although the team will consider their support requirements when researching a tool, the ultimate decision is around whether they will be locked into a tool and what the administration effort is like. However, neither of these concerns is effective in picking the right tool as lock-in and administration are both subjective and time-dependent. If a team implements a tool in the wrong way, then the administration effort will be higher than those that implement the tool correctly. And if the team solely focuses on building their processes around what the tool does, then they will be locked into that tool. The best way to approach tool selection and implementation is by understanding what the best practices for the process are.

Best practices, which are practices that are the most effective, exist for almost every tool, process, or implementation. By understanding the best practices for the underlying process that a tool is looking to integrate with, the lock-in and administration concerns become secondary criteria. For example, if a company wants to implement Kubernetes, they might create a cluster in a cloud provider and deploy to it to see how it works. The decision of which cloud provider to host Kubernetes on will inevitably be made based on potential cloud provider lock-in and the related administration requirements. However, if the company understands the best practices around implementing, configuring, and operating Kubernetes before they pick a hosting solution, they will find that the cloud provider doesn't matter as much. If they implement Kubernetes best practices in general and then choose a cloud provider to host their cluster, they can easily move to another cloud provider whenever they want. This ease of migration is the result of deciding on best practices first and then picking a provider that meets those best practices. And since every cloud provider has a similar underlying infrastructure, the Kubernetes cluster migration will be significantly easier.

In regard to continuous deployment, delivery, and GitOps, it is important to break down what the best practices are, how the company will abide by those best practices, and then what functionality or support must exist for those best practices to be met.

Continuous deployment considerations

The DevOps team began breaking down the different parts of their deployment process. They also started gathering complaints and concerns from developers, and figuring out what industry practices should be implemented to achieve continuous deployment.

One of the major concerns that they had heard from the different teams was that every deployment was a black box for them. From the moment the deployment was triggered, the team had no understanding or insight into how it was behaving. A failed deployment could take hours to find and fix. Although the resolution to this would be increasing the log output, the core issue is the lack of transparency. The ability to achieve a continuous deployment practice relies heavily on transparency into what is happening and shortening the overall feedback loop to the teams. The DevOps team would need to implement transparency and feedback as a best practice for deployments.

Another part of the current deployment process that causes some concern is the reliance on multiple different teams and tools to get a release candidate ready. The service-oriented architecture requirements mean that every team is dependent on the slowest-performing team in the group. To allow their teams to achieve a continuous method of deployment, they would have to decouple the microservice deployments and allow them to deploy automatically. This requirement will result in another best practice: that a microservice will be automatically deployed when it gets built.

One last area that many teams in the industry are implementing is how the deployments are verified as they go out to production. Although the team could simply implement a health check command, the desired verification step would need to consist of multiple types of checks against different metrics and error messages. By having an easy way of verifying the deployment, the team would have more confidence in frequent releases. Also, if the verification checks not only happened in production, but also in pre-production, the team could avoid issues from progressing to higher-level environments. What the deployment process would need is a more intelligent or easy-to-use verification step, especially for notifications or dashboarding.

The best practice list for the team now consists of deployment transparency, deployment feedback loop, automated deployment triggering, and deployment verification. Now, the DevOps team has to consider how they can implement these best practices in Ansible or Harness.

Continuous deployment, as a process, is something that many engineering teams hope to adopt. The promise of smaller changes, fewer bugs, and shorter lead times are all enticing. And yet many teams do not achieve this desired outcome because they look to the end goal of daily deployments, and they are not sure what they need to change to achieve that frequency. Some assume that reducing the size of the teams will allow for more frequent deployments. Others think that reducing the size of the services that are deployed will allow for reduced lead time. But in many cases, the lack of confidence in the deployment process or verification of production are significant barriers to even pursuing a continuous deployment frequency.

When defining build best practices for continuous deployments, it can appear that the easiest place to start is understanding what prevents a team from achieving daily deployments in the current method. However, building best practices based on current issues is not actually getting to a set of best practices, but rather compiling a set of best current practices. The current best practices are entirely subjective, limited to the scope of the current process and tools, and will change shortly after they are made and implemented. Continuous deployment best practices should have a more objective definition and withstand the changing landscape of tools, architectures, and platforms. The following best practices, although not an exhaustive list, should be considered for any continuous deployment process:

1. **Transparent Deployments**: Regardless of the platform, architecture, or tool, every deployment should allow for sufficient transparency. Any person observing the deployment should be able to understand where in the deployment process the current execution is, details around what has already occurred, what is currently happening, and what is left for the deployment to be completed. Each part of the deployment process needs to also include transparency at a level where adequate feedback into issues or potential issues is easily understood. This includes detailed logging, graphical health representation, and immediate error recognition.

2. **Loose Coupling of Dependencies**: As other teams, tools, platforms, and architectures change, so do dependencies on those different verticals. The deployment process should reduce or alleviate any dependencies that would prevent a service from being deployed. Some dependencies, such as the endpoint where the artifact is deployed, are required. But even with those requirements, the deployment process should operate without any barriers to entry. Every service should have little to no entry fee when being deployed.

3. **Confident Deployments**: Every deployment needs to garner the utmost confidence from the deployer and the monitors. Confidence is built through two main avenues: **testing** and **verification**. The testing process should be partially built by the developer of the code, a quality assurance team, and a production monitoring team. Each test should cover the functionality that is built, its interaction with other functions that it interacts with, and any general functionality that relates to business metrics. The tests need to be automatically executed as the deployment process proceeds and should be frequently improved. Verification is the other requirement for increased confidence in a deployment. Every well-tested service still requires a verification process in the higher-level environments. Verification should consist of monitoring hardware metrics, user metrics, and business metrics, as well as logging verification associated with code execution. By implementing effective verification with broad coverage, the confidence that an issue will be found and remedied before a customer discovers the problem will increase.

These best practices will assist in building out an effective continuous deployment process that will increase deployment frequency, shorten developer lead time, reduce mean-time-to-restore, and even have a positive impact on the change-failure rate of the applications. But a true continuous deployment process wouldn't be truly continuous without a continuous delivery process surrounding it.

Continuous delivery considerations

As the DevOps team deliberated over what deployment best practices would be like for their company, they realized that they would also need best practices for before and after the deployment was done. Every developer that had releasable code would need to pass their artifacts through a set of testing sequences. But before they could submit the artifact for a deployment, they would have to work through a set of change management practices to ensure that sufficient approvals and documentation were submitted. Because of the company's compliance standards, the artifact would not be able to be deployed to production by the same person who submitted the code. This requirement created at least a two-person release process. Additionally, audit-related requirements meant that significant access controls needed to be in place, and changes that had been made to any service needed to be documented. Every release would also require a set of documented and automated verification and failure remediation steps, such as rollbacks, with root cause analysis practices as a part of the failure remediation requirements. Although all of the pre-deployment and post-deployment requirements were well understood, the best practices around these requirements needed to be better defined.

The first of the best practices is around change management. Although the requirement of having proper approvals and documentation was well-understood, relying on a person or set of people to manually work through the process will always result in bottlenecks. Instead, the best practice should be that any change management steps should be repeatable because of automation, reliable because of sequential execution, and scalable because of templatization. Every delivery pipeline that's executed should provide the relevant information to a command that automatically generates the appropriate change management document, and then waits for the desired outcome before the next step is allowed to move on. The DevOps team will need to implement some level of dependencies to prevent anyone from skipping these requirements as well.

Another best practice for continuous delivery is automated failure strategies. The deployment process should allow for easily detecting the status of the deployment and executing the appropriate failure strategy, based on what has already been successfully accomplished. If there was an issue with the creation of a change management ticket, then the failure should just stop the pipeline, as there is nothing to roll back to. However, if the deployment was taking too long or there was a noticeable error, a rollback process should be executed. Helm would allow for an easy rollback execution because of its built-in rollback commands.

The last best practice for the DevOps team to implement would be environment configuration consistencies. Every deployment that goes out should be deployed in the same way with the same environment configuration steps. By making the environments consistent and immutable in their configuration, every deployment has inherent repeatability and reliability. The only differences regarding the environments should be the names and network requirements, but the configuration of the environments should be consistent.

Through enforced and automated change management, automated failure strategies, and environment consistencies, the DevOps team can ensure that these continuous delivery best practices will be applied to any tool that they choose to bring in. Ansible and Harness both have the ability to execute commands in sequence, to enforce failure strategies based on dependencies, and to provide values to the different templates that are executed. These best practices, coupled with the continuous deployment best practices, will enable the team to accomplish their desired continuous outcome without needing to rely on any specific tool, architecture, or platform.

Within continuous delivery, there are different stages that much be executed. Each stage is a segregation of work, typically covering pre-deployment, deployment, and post-deployment. Best practices for each stage become an essential part of a successful delivery pipeline. Delivery is often under-designed because there is a lack of collaboration across all teams involved in the life of a delivery execution. Developers may be the ones writing the code, but the quality, security, infrastructure, DevOps, and SRE teams are all involved in a successful delivery pipeline.

Every company is different in terms of their requirements for a delivery pipeline as well. Some companies have federal requirements that prevent some users from deploying to production. Others require heightened data protection for health code certification. And then there are some companies that require advanced compliance and auditing requirements because of finance-based auditing and certification. But what is important to note about these different certification and auditing requirements is that best practices are the same across each of them.

The best practices for deployments that were previously discussed are transparency, loosely coupling dependencies, and confidence-building practices. The continuous delivery best practices to add to the list are as follows:

1. **Change Management Automation**: Every environment and every deployment requires some level of change to occur. That change can range from building an entirely new environment from the ground up to a change in a variable value for a service to use. Although documenting what the changes are is important, enforcing the change to happen and checking that it did happen is essential to success. If the desired execution step is to halt a process and wait for a manual interaction on the desired environment before moving on with the process, a bottleneck will inevitably occur. By adding automation that makes these changes and then checks the state of the change, the change is enforced, and the bottleneck is avoided.

2. **Failure Strategy Automation**: As the deployment frequency increases, the number of deployment failures will also increase. Although the pursuit of reducing failures is desirable, reducing service interruption is more important for the company. Having a strategy for how to detect a failure and what to do when a failure is detected is known as a **failure strategy**. Detecting a failure is something that can be done by running smoke tests, using a log aggregation tool, using a performance monitoring tool, employing user acceptance testing, and even waiting for a user to report an issue. Each of these failure detection methods will determine the severity of the issue, what testing should be improved to detect the issue sooner, and what actions should be carried out to resolve the failure. If the failure occurs during a deployment, a simple rollback can solve the problem before the service is interrupted. The same is true of a smoke test, where a rollback can solve the problem before service interruption. With regard to log aggregation and performance monitoring, the engineering team that either deployed the service or is responsible for service availability will need to be alerted as soon as possible to any issues. That alert should show what error is occurring, if the service is being interrupted, if users are being impacted, and offer enough troubleshooting information to dictate what actions should be taken to restore the service. The last failure detection method is relying on the user to voice a concern. This method is the least desirable as it indicates that at least one user is being negatively impacted by the failure. As such, the failure should be remedied as quickly as possible before other users encounter the issue and before the business is negatively impacted.

3. **Consistent Environment Configuration**: If there are multiple environments and each environment configuration is different, then every deployment becomes different from every other one. However, if the differences between how the environments are configured include the name of the environment and some of the network specifications, then each deployment is essentially the same. By reducing the complexities from one environment to the next, the reliability, repeatability, and scalability of the deployments grow even as the number of services scales.

Although these six best practices might not be an exhaustive list, they are an essential list to start with. By understanding and implementing these best practices, the engineering teams can ensure that they will not be locked into a tool or vendor and that the overall administration will be lighter for their teams. In some cases, they might find a tool or solution that has more features or functionality than others in the market, but those additions should only work to aid in implementing the best practices. Also, whether the tools allow for more automation and less scripting or not, the most administrative work to accomplish these best practices will be on the front of the implementation, rather than throughout the life of the tool's use. The last piece to consider is how to use GitOps to adopt and enforce these best practices.

Where GitOps ties in

The DevOps team now have a list of best practices that they want to implement for their delivery and deployment processes. They know that they will need to provide their users with deployment transparency, they will need to loosely couple the deployment dependencies, and that confidence in the deployment is essential to frequent deployments. They also know that all the changes should be automatically applied and checked, failures need to be easily detected and remediated, and that the different environments should be as consistent as possible. But what they want to figure out is how to bring all of these best practices together with GitOps.

The best thing about leveraging GitOps for their delivery and deployments, whether that is with Ansible or Harness, is the native reusability of the code. Because every pre-deployment step, deployment step, and post-deployment step was defined in code, they could enforce consistency at each point in the process. The next step for the DevOps team was to define each of the best practices in code.

Some of the best practices would be easy to enforce as they were settings associated with the tools being used. To allow for maximum transparency, they need to break out each step into smaller steps that would report back the execution logs at a more detailed level. If they could make the logs accessible to their end users, then the feedback loop would be self-service.

Another easy to enforce best practice is environment consistency. They would work with the cloud infrastructure team to build every environment so that they're as close to each other as possible, and then require the same variables to be leveraged for every environment. Although some of the ancillary services, such as databases and message queues, would be different for each environment, leveraging the same name for the location and connection variable would make each deployment consistent.

The other best practices will require more work for the DevOps team to enforce. Loosely coupling dependencies is partially dependent on the culture of the different teams and partially on the ease of deployment triggering. If each developer can get their changes into production in an automated way, especially if that trigger is based on the artifact being built, then dependency decoupling will naturally occur.

Building in change management automation will be simple, since the configuration changes can be defined in a set of configuration files that are consumed by the delivery pipeline. The developer would have a set of configuration variables that are predefined, and the DevOps team would provide a set of environment configuration files that would also be predefined. Because GitOps is a declaratively defined process, the configuration files can live in the same code repository as the application code and the delivery pipeline code.

Building confidence and detecting failures can be tackled in the same set of steps. Smoke tests and action steps for a failure can be defined in code for the delivery pipeline. The only thing that the DevOps team would need to consider is how to execute the different failure strategy steps based on the outcome of the smoke tests, the monitoring tools, or user interaction.

Once the DevOps team was able to get a list of the best practices and an understanding as to how the best practices should be enforced, they were able to better comprehend how to represent the best practices in code. Ansible would require more work to implement the best practices than Harness would, but both tools would allow for a GitOps method of best practice implementation.

In every style of GitOps, each step is intended to be declaratively defined. As the breadth of the GitOps coverage increases, so does the amount and size of the declarative file requirements. One of the advantages of leveraging GitOps is that the inherent declarative file requirements automatically result in accomplishing some of the best practices by default. For example, the best practice of environment consistency means that the GitOps process requires fewer declarative files. This is because there are minimal significant changes between each environment to represent in code. If the only difference between the environments is the information that's provided to a variable, then fewer declarative files are ultimately required.

However, depending on the tool that is being used, some of the delivery pipeline configuration files will vary in regard to the size and number of requirements to accomplish the same task. For example, a tool that has a native integration with a change management tool will allow for fewer delivery pipeline configuration files than a tool that does not have the same integration. If an integration is lacking, then the team implementing the tool will be required to build the integration scriptand associated files.

Regardless of the tool that's chosen, the integrations that are available, or the amount of work required to implement best practices in GitOps, the outcome will almost always be worth the effort. What is important to understand for any team embarking on the journey of enforcing continuous delivery and deployment best practices through GitOps is that time to value is more important than cost saved. Every engineering team has enough skill to build an entire GitOps tool. However, every engineering team needs to consider what their company's competency is. If the company's competency is not in continuous delivery, continuous deployment, or GitOps, then there should only be a few, if any, engineers associated with building, implementing, and maintaining a GitOps tool.

Summary

This chapter was focused on exploring the purpose of best practices, the different best practices for continuous delivery and deployment, and how GitOps ties into those best practices. We spent some time learning about how the DevOps team in our analogy considers the different best practices, how they discovered what best practices were, and how they would need to implement them.

In the next chapter, we will work through the basics of building different declarative files in different languages. You will learn how to structure JSON, YAML, and XML, as well as understand the different templatization styles that can be used. The following chapters will be more technical in nature, but there will be significant explanations along the way.

Section 3:
Hands-On
Practical GitOps

Now that you have been provided with an overview of GitOps and the fundamentals have been covered, in this section, you will learn what your current state is and how to get to a desired state of continuous delivery and continuous deployment.

This section comprises the following chapters:

8
Practicing the Basics – Declarative Language File Building

Before getting into the actual work of implementing a GitOps practice or tool, it is important to know about how a declarative file is built, structural syntax, different language types, and templatizing capabilities. Once the basics are understood, a team can build upon a standard that will be shared across a company.

An important thing to understand about choosing a declarative language to standardize on is that a company or team doesn't have to standardize on one language. Some tools have the ability to leverage multiple declarative language types, and if a solution is chosen that only supports one declarative language type, then there are tools that exist that can help convert declarative languages from one type to another.

In addition to tools that leverage declarative languages, there are also platforms that leverage declarative languages. For example, Kubernetes is well known for using YAML, and nearly every service in AWS uses JSON. As such, it is important to understand what is and is not supported to avoid extra steps in the deployment and delivery process.

In this chapter, we're going to cover the following main topics:

- Nesting and flat files
- XML file building
- JSON file building
- YAML file building
- Templatization types

Nesting and flat files

The DevOps team was still waiting on leadership to decide on the tool that would be used for their verified GitOps practice. The decision was to be made relatively soon, but the more time that the team waited to get started on the project, especially with a tool that might require more time to set up and configure, the fewer required functionalities would exist on the initial launch of the tool.

The team decided that the best use of time would be to begin training their engineers on how to leverage declarative language files for as many parts of their processes as possible. The development team should be able to provide a list of variables in a configuration file for the deployment to use. The quality team would need to build their testing to leverage configuration files as well, making the testing scalable across all microservices. The cloud infrastructure team was already leveraging Terraform, which has its own declarative language file support, but it would be helpful if some of the commonly requested tasks could be triggered and configured by those outside of the cloud infrastructure team. These were relatively lofty goals for the DevOps team to implement for the engineering group, but it would be much easier to implement the desired repeatability, reliability, and scalability of the many different operations across all teams.

As the DevOps team started to work on the material for training the engineering teams, they realized that the biggest issue they needed to tackle was the structure of the files. They needed to define a way to separate duplicate values, whether the duplicates exist in the same team or across multiple teams, how to store the files, and how to reference the files. Then, once that decision was made, they would need to show the teams how best to build out a declarative file or use a pre-made file that the DevOps team can provide for them.

When approaching a new programming language, the fundamental concept to understand is the syntax, or the required arrangement of words and phrases. Some languages have a syntax that requires certain characters to mark the end of lines, the beginning of loops, or different types of variables. The syntax of a declarative language file is also important to understand, but the syntax is relative to the desired adoption method of the files. The core makeup of a declarative file is not the syntax of the language, but rather the structure of the file being either flat or nested.

A **flat file** structure is when every variable or every grouping of variables is aligned to the left of the file without tab or space indentation.

Example 1.1: Flat file basic structure:

```
# BASIC STRUCTURE
name-1: "hello-world-1"
description-1: "say hello to the world 1"
count-1: 1

2-name: "hello-world-2"
2-description: "say hello to the world 2"
2-count: 2

nameThree: "hello-world-3"
descriptionThree: "saying hello to the world 3"
countThree: 3
```

Example 1.2: Flat file grouping structure:

```
# GROUPING STRUCTURE
[GROUP 1]
name: "hello-group1"
description: "saying hello to group 1"
count: 1

[GROUP 2]
name: "hello-group2"
description: "saying hello to group 2"
count: 2
```

```
[GROUP 3]
name: "hello-group3"
description: "saying hello to group 3"
count: 3
```

This file breaks down into two main sections: the basic structure in *Example 1.1* and the grouping structure in *Example 1.2*. Both of these sections are used in different scenarios, but the purpose of using the different types of flat file structuring is to mitigate issues with duplicate variables names. In *Example 1.1*, the potential duplicate variables are separated by leveraging a suffix, a prefix, or a naming style (examples of naming style include camelCase, kebab-case, and snake_case).

The alternative to leveraging a naming convention is to leverage a grouping structure, which uses a certain syntax to define a specific set of variables to target by declaring which variable and group to reference. In *Example 1.2*, the grouping structure is separated based on the name of the group, which is surrounded by [and] characters. A common file that uses these bracket characters is an initialize file, also known as an `ini` file, which is a popular style in Windows applications.

Although the readability of the flat file structure is relatively simple, the complexity increases relative to the size and number of variables contained in the file. If a user is leveraging less than 20 unique variables, then a flat file might be the easiest structure type to use. However, if the files are intended to be longer or used across multiple teams, platforms, or tools, then a nested file structure would be the most user-friendly and readable.

A **nested file**, as the name suggests, is a file that nests variables into groups and subgroups by leveraging indentation with tabs or spaces.

Example 1.3: Nested file structure:

```
group-1:
  name: "hello-world"
  description: "say hello to the world"
  count: 1

group-2:
  name: "hello-world"
  description: "say hello to the world"
  specifics:
    count: 2
```

```
    label: "app"


group-3:
  name: "hello-world"
  description: "saying hello to the world"
  specifics:
    labels:
      name: "app"
      count: 3
      secrets:
        name: "secret-name"
        value: "secret-value"
```

The nesting structure in *Example 1.3* is self-explanatory, in that there are **groups** and **subgroups** defined by nesting relative variables together:

- The first nested group in the preceding file is a basic nesting group that has a set of variables defined at the same level, or indentation, in one group.

- The second nested group contains a group and subgroup that are set at two different levels, meaning that within the parent group of group-2 there is a child group called specifics.

- The third nested group contains a group and a set of subgroups. The breakdown of this third group is a parent group called group-3, a child group called specifics, a grandchild group called labels, and a great-grandchild group called secrets.

The group names in the nested structure can either be teams, tools, platforms, environments, and so on. But the goal of the nested structuring is to allow for an easy-to-understand breakdown of variables. One last piece to understand about declarative files, specifically with nested structuring, is the different types of objects that are available for use.

Another part of building out declarative files is understanding the different types of variables and the different types of objects that can be used. Variables are essentially pairs of keys and values that allow different kinds of data to be assigned and used. The most common variable types are string, integer, and boolean.

Example 1.4: Variable type examples:

```
string: "This is a string"
integer: 1
boolean: True
```

In *Example 1.4*, `string` is declared using double or single quotes, `integer` is a number, and `boolean` is either `True` or `False`. These different variable types are common across all declarative languages allowing for easy readability and usability.

However, there are some significant differences between the different declarative language types in regard to objects. `object` is a type of styling for data that is leveraged for different reference purposes, which will be explained further in the JSON section of this chapter. But, for now, it is important to understand the basic object types that exist and why they are used. These object types are `object` and `array`.

Example 1.5: Object type examples:

```
object: {name: "John", age: 30, male: True}

nested-objects: {name: "Jane", age: 30, address: {city: "New
York", state: "New York"}, female: True}

array: ["first", "second", "third", 4, 5]

nested-arrays: ["first", "second", ["third", 4, 5], 6, "seven"]

array-with-object: ["first", "second", {name: "third", age: 4},
5]

object-with-array: {name: "Duke", age: 20, favorite-foods:
["pizza", "ice cream", "fish"]}
```

Example 1.5 shows both the basic structure of an object type and also that object types can be nested and cross-nested. An object gives the user a way to group a set of keys and values together for specific reference later. An array, which is a collection of elements, gives the user a way to detail a list of values that are grouped together, but without requiring a key or name to assign values to. Objects can contain pairs of keys and values, other objects, and arrays, whereas arrays contain mainly values but may also contain other arrays and objects. Understanding the distinctions between the different variable and object types will allow easy readability and usability across different platforms, tools, and processes.

With the different declarative language fundamentals shown so far, the next part of the process is understanding the different declarative language types and then choosing one as the main standard.

XML file building

While building out examples for the enablement presentations, the DevOps team needed an easy way to explain all of the different variable and object types that a user might encounter. Although all of the developers would have a working knowledge of the different types and uses, they were not familiar with leveraging declarative language files to define how the deployment and delivery process would behave. For the other teams in the engineering group, the use of declarative language files was few and far between. Some of the database administration team was familiar with NoSQL data structures that used declarative language structure, such as JSON, for data storage. The quality assurance team had mainly used declarative files for some of their tools, but the files required little editing outside of the default settings. The team with the most familiarity in declarative language file building was the cloud infrastructure team, since they were using Terraform. But even those users that were deploying to Kubernetes were leveraging Helm charts that were built for them, rather than building or editing the charts themselves.

The DevOps team needed to find a basic declarative language that allowed the enablement to start at the fundamentals and then move on to a more advanced declarative language as the training progressed. And since everyone in engineering was already familiar with the basics of HTML and website creation, the easiest declarative language to start with was XML.

XML uses the same open and close tagging style that HTML uses, since it is also a markup language like HTML. But where XML does not have the different variable or object types that JSON or YAML would, it does have the ability to associate attributes to the tags. The DevOps team would need to leverage these attributes to assist with the learning transition to JSON and YAML.

XML, which stands for **Extensible Markup Language,** is a declarative language type that was designed to store and exchange data. It is still commonly used today in web applications, some IT systems, and any other data representation use cases. But a major difference between XML and other declarative language types is that XML is designed to be self-descriptive.

Example 2.1: XML file example:

```xml
<grocery>
  <food category="fruit">
    <name>Strawberry</name>
    <color>Red</color>
    <price>1.99</price>
    <package>1 lb</package>
  </food>
  <food category="vegetable">
    <name>Carrot</name>
    <color>Orange</color>
    <price>3.00</price>
    <package>3 lb</package>
  </food>
  <food category="meat">
    <name>Ground Beef</name>
    <color>Red</color>
    <price>2.99</price>
    <package>1 lb</package>
  </food>
  <food category="grain">
    <name>Brown Rice</name>
    <color>Brown</color>
    <price>0.99</price>
    <package>1 lb</package>
  </food>
</grocery>
```

Example 2.1 shows a basic XML document that describes a list of grocery items. The idea of XML being self-descriptive means that there are no predefined tags that must exist in the XML document, unlike HTML. HTML uses tags such as `<p>`, `<h1>`, and `<body>`, whereas XML uses any tag that makes sense for the document. Every group is defined with indentation, which creates a nest of related items, and is defined with a category attribute. The self-descriptive behavior of XML allows any solution to pull only the information that it knows should exist. For example, if a tag was inserted into the XML document that a solution did not know about, the solution wouldn't pull any of the information from that tag. There are some additional benefits to the self-descriptive structure of an XML file, especially when using advanced structuring.

Example 2.2: XML file declaring the attributes of a Honda Civic:

```
<car category="Sedan" risk="low">
  <make>Honda</make>
  <model>Civic</model>
  <year owner="1">2020</year>
  <mileage usage="own">15,000</mileage>
  <technology>
    <feature repaired="false">Backup Camera</feature>
    <feature repaired="true">Power Seats</feature>
  </technology>
</car>
```

Example 2.3: XML file declaring the attributes of a Nissan Armada:

```
<car category="SUV" risk="low">
  <make>Nissan</make>
  <model>Armada</model>
  <year owner="2">2018</year>
  <mileage usage="lease">30,000</mileage>
  <technology>
    <feature repaired="false">Backup Camera</feature>
    <feature repaired="true">Power Seats</feature>
  </technology>
</car>
```

Example 2.4: XML file declaring the attributes of a Toyota Prius:

```
<car category="Sedan" risk="medium">
  <make>Toyota</make>
  <model>Prius</model>
  <color>Silver</color>
  <year owner="1">2014</year>
  <mileage usage="fleet">90,000</mileage>
  <technology>
    <feature repaired="false">Backup Camera</feature>
    <feature repaired="false">Power Seats</feature>
  </technology>
</car>
```

Example 2.5: XML file declaring the attributes of a Chrysler Town and Country:

```
<car category="Minivan" risk="high">
  <make>Chrysler</make>
  <model>Town and Country</model>
  <color>Beige</color>
  <year owner="4">2010</year>
  <mileage usage="own">125,000</mileage>
  <technology>
    <feature repaired="false">Backup Camera</feature>
    <feature repaired="false">Power Doors</feature>
  </technology>
</car>
```

The advanced structuring shown in *Examples 2.2* through *2.5* allows the user to search for attributes and pull any information related to them. For example, if the user wanted to get a list of all sedans, they could search for `category="Sedan"` and pull back information related to the Prius and Civic. The user could also use a search to find any groups with low risk, similar to the search for the sedans.

The XML structure allows a user to quickly understand what is meant by a declarative language file. This ease of understanding is because of how human-readable the XML structure is. All tags and values are declared by the user and can be used by a tool or solution to alter the behavior of the execution. However, XML can become very verbose and difficult to scale, since it does not allow templating. JSON and YAML are the most common declarative languages for delivery and deployment tools.

JSON file building

With the cloud infrastructure team using Terraform already, the DevOps team would be able to leverage some of their expertise in enabling the rest of the engineering organization on the structure of JSON. Although the XML training would be easy and lightweight, the JSON training would have an effect on the actual tooling for the delivery and deployment processes. The DevOps team had also planned on giving the engineers the ability to leverage either JSON or YAML for their configuration files, allowing the tool to do the conversion and application work for them. Currently, Ansible and Harness do not have this capability, meaning that the DevOps team would need to build or leverage an intermediate tool that converts the files into the desired language types for native consumption.

The first step in the JSON training is deciding how the different teams will build out their files, whether each team will have their own or one main file will be shared across a whole team. Then, the DevOps team will need to show the engineers how to leverage objects and arrays in the JSON file to detail the desired outcome of the execution. A generic file will need to be built by the DevOps team with sufficient notes for the engineers to manipulate the files in the correct way. The main issue that the DevOps team would face is designing the tools to consume the files in a certain way and ensuring the integrity of the file structure and the provided values.

JSON, which stands for **JavaScript Object Notation**, is similar to XML in that it is used for storing and exchanging data. But the main difference is that JSON is structured closer to a programming language than XML is, using objects and variable types that are predefined, but allowing a similar self-describing style as XML.

A major advantage of JSON over XML is that JSON doesn't require open and close tags, but rather uses characters such as { } and [] with comma delimiting between different variables and objects.

Example 3.1: JSON file declaring customer attributes:

```json
{
  "customers": [
    {
      "name": "John Smith",
      "location": {
        "state": "NY",
        "city": "New York"
      },
      "age": 30,
      "company": "JSmith LLC"
```

```
    },
    {
      "name": "Jane Smith",
      "location": {
        "state": "CA",
        "city": "Los Angeles"
      },
      "age": 25,
      "company": "Jane and Co"
    }
  ]
}
```

The JSON in *Example 3.1* shows a breakdown of the `"customers"` array into individual objects that define each customer and any required information about them. There are nested objects in this file as well to assist with easy data retrieval. If the `"location"` object was an array, there would be no way to reference only the city or state of the customer. However, the array of customers allows a reference to all objects in the array; or, using the index location of the objects in the array, a user or tool can reference a specific object.

For example, if a user wanted to get the second object in the array, they would use `customers[1]`, which would give them the second item in the `"customers"` array (computers start their counting at `0`, which is why the index of `1` is the second item in the list). However, if a user wanted to get the state of the second customer, they would use `customers[1].location.state` to get the information from the nested object. This is called dot notation and breaks down to `array[index].object-name.variable-name`. The use of dot notation is very common when leveraging any type of JSON structure and some tools will output information in a JSON structure to make it easy for a user to use the data from the output.

A common use case for JSON in regard to a tool or platform would be AWS's ECS service. **ECS**, or **Elastic Container Service**, is a service that is similar to Kubernetes in that it allows for easy container orchestration, but without all of the administrative overhead of Kubernetes being a requirement. In ECS, every container is called a **task instance** and is defined using a **task definition**, which is a JSON representation of what kind of task the user wants in their ECS cluster.

Example 3.2: JSON file declaring an ECS task definition:

```json
{
    "family": "",
    "taskRoleArn": "",
    "executionRoleArn": "",
    "networkMode": "",
    "containerDefinitions": [{
            "name": "",
            "image": "",
            "cpu": 0,
            "memory": 0,
            "logConfiguration": {},
            "firelensConfiguration": {}
        }],
    "volumes": [{}],
    "cpu": "",
    "memory": "",
    "tags": [{}]
}
```

The task definition in *Example 3.2* has certain predefined requirements that AWS is expecting when this file is being used. As shown, there is no documentation in the file that states what is and is not required, which means that a user must read through the documentation from AWS to understand what values are required and which ones are not.

A task definition serves the purpose of providing AWS with the desired end state of this particular task. The container image used, the networking requirements, any environment settings, and so on are all defined in this file. When the file is provided to AWS, there is an underlying execution engine that AWS has that will take the information from the file, add it to the execution configuration, and then the ECS cluster will be directed to achieve the declared outcome. Similar to how AWS ECS achieves this process and leverages this task definition file, GitOps-based deployment and delivery tools will use a similar execution engine process that leverages these file types.

In regard to ECS and Kubernetes, which use manifests or task definition files, the execution engine may pass the manifest or definition file from a Git repository to the platform endpoint. The instructions to get the file from the repository, which platform and infrastructure to direct the file toward, and any before- and after-deployment steps will also be represented in a declarative file. This combination of files can be understood as configuration code, which are the declarative files associated with the pipeline execution configuration, infrastructure code, which are the declarative files that need to be passed to the infrastructure tool or platform, and pipeline code, which is the declarative files associated with the execution engine pipeline configurations.

JSON is one of the most common ways to handle data output and data retrieval, especially when running health check processes, almost any service in the cloud, and most infrastructure-as-code tools. But there are some widely used platforms and tools that leverage YAML files for configuration and manifest building.

YAML file building

Engineering leadership has decided that the platforms that need to be supported by any delivery or deployment process must include containerized, serverless, and traditional applications. Each of these different platforms has its own structure and requirements, which means that a one-size-fits-all solution is difficult to implement. Some of the platforms and cloud providers that the company is looking to leverage require JSON-based configuration files. However, one of the main platforms that the teams have been leveraging and will continue to leverage is Kubernetes, which is mainly configured through YAML manifests.

Since the development teams have already been deploying via Helm charts, which use YAML manifests, there is some familiarity with the YAML structure already. However, the quality and infrastructure teams mainly use JSON for their data storage and exchange, as well as for some of their tooling, such as Terraform. Because of the differences between the JSON and YAML structures, the enablement portion around YAML will be focused on the general style of a YAML file and notable differences. Both JSON and YAML are easily readable based on the syntax, but the main issue will be around the different object types and uses. The most noticeable difference is the use of different characters that wrap around objects and variables to denote types and line endings. Some of these character uses are the same in YAML, but YAML will also use indentation and newlines as a part of its syntax.

YAML, which stands for **YAML Ain't Markup Language**, is similar to XML and JSON in that it is used for data storage and exchange. YAML is also self-descriptive, allows the nesting of objects, and is declarative in nature. Where YAML differs is in the character usage and lack of end-of-line delimiters. XML uses tags to denote the beginning and end of a line, JSON uses either characters or commas to denote the beginning and end of a line, but YAML mainly relies on indentation and new lines to indicate the beginning and end of a line.

Example 4.1: Comparison file:

```
<!-- XML STRUCTURE-->
<key>value</key>


// JSON STRUCTURE
{"key": "value"}


# YAML STRUCTURE
key: "value"
```

Example 4.1 shows how a key and value pair would be represented in the different languages. XML has the key located in the tags, which are wrapped around the value. Although JSON would typically just define the key and value as strings with quotes around them, a JSON file will typically start and end with { } to make the entire file contents a single object. YAML requires at least one key and value pair, with the value type being defined by its representation in the code.

YAML is commonly leveraged when referring to Kubernetes or a Kubernetes-related tool, such as `helm`, `kustomize`, `istio`, and others. But YAML is not typically used by applications for data output, as data in YAML files is often accessed using dot notation and specific built-in templating engines.

Another between JSON and YAML is the use of object types, such as arrays. Although YAML doesn't have an object structure as JSON does, there are certain characters or indentations to represent different object types, such as **scalars**, **maps**, and **sequences**.

Example 4.2: YAML object type examples:

```
# SCALAR
name: "John Smith"


# MAP
address:
  city: "New York"
  state: "NY"
  phone:
    area-code: 123
    phone-number: 4567890
```

```
# SEQUENCE
cars:
  - make: "Honda"
    model: "Civic"
    color: "Black"
    mileage: 100000
  - make: "Toyota"
    model: "Corolla"
    color: "Silver"
    mileage: 85000
```

Example 4.2 shows a scalar, a map, and a sequence. A scalar is just a key and value pair, a map is a nested grouping of scalars, and a sequence is a parent object that contains multiple groups of scalars that are denoted using the – character. The sequence is the most unique object type in YAML, but it would be similar to an array of objects in JSON.

A real-world example of scalars, maps, and sequences in YAML would be building out a Kubernetes manifest for a deployment.

Example 4.3: Deployment YAML file:

```
apiVersion: apps/v1
kind: Deployment
metadata:
  name: app
  namespace: default
spec:
  selector:
    matchLabels:
      app: app
  template:
    metadata:
      labels:
        app: app
    spec:
      containers:
      - name: app
        image: library/nginx:latest
```

```
    resources:
      limits:
        memory: "128Mi"
        cpu: "500m"
    ports:
    - containerPort: 8080
  - name: sidecar
    image: library/istio:latest
    resources:
      limits:
        memory: "128Mi"
        cpu: "500m"
```

In *Example 4.3*, the declaration of apiVersion is a scalar, metadata is a map, and the containers section is a sequence. These are all used for different reasons, which are predefined in Kubernetes, and are based on how the execution engine of Kubernetes intends to interpret this information. Some of these requirements from the underlying platform make the YAML file less self-descriptive, since there are reserved words and required datasets. But when using YAML for other tools, YAML can be significantly more self-descriptive. For example, in regard to Helm, there is typically a resource file in a chart and a values file to pass variables into the resource file. The resource file will be structured similarly to a Kubernetes resource file, but with indicators for variables. The values file is used to pass in values to the resource file, where the variables are found.

Example 4.4: Pod YAML file with go-templating, often found in Helm charts:

```
# POD YAML
apiVersion: v1
kind: Pod
metadata:
  name: {{ .Values.name }}
spec:
  containers:
  - name: {{ .Values.name }}
    image: {{ .Values.image }}
```

Example 4.5: Values YAML file:

```
# VALUES FILE
name: app
image: library/nginx:latest
```

Example 4.5 is a values file that is self-descriptive, since it uses a user-defined set of scalars. The values file often exists at the root of the `Chart` folder, next to the `Chart.yaml` file. Whereas the `Pod` file in *Example 4.4* does require a predefined set of information from Kubernetes, the reference to the values file is done through dot notation and leveraging the self-described scalars in the values file. The `Pod` resource file will often exist with other resources files in a templates folder at the root of the `Chart` folder.

Although XML, JSON, and YAML all provide ways for data to be stored and exchanged, there are significant advantages to leveraging JSON and YAML for the delivery and deployment process. This is especially true when considering that delivery and deployment tools, as well as the underlying platforms and solutions, all use either JSON or YAML for configuration. But, if the user was required to make a set of static declarative files, the potential for manifest sprawl would be great. In order to make the administration burden easier for a DevOps team, the teams will often resort to templating options with JSON or YAML files.

Templatization types

Once the DevOps team had a better understanding of the different platform and tool requirements, they found that the easiest method forward was to give the engineers minimal configuration files. If the main files for both JSON and YAML were built and maintained by the DevOps team, then the engineers would only have to provide a small subset of configuration variables. The delivery pipeline could be built to get the configuration file from a Git repository and then pass in the values from a predefined set of variables.

But to make the DevOps team's administration effort as light as possible, they would need to implement a type of templating process for JSON and YAML files. This way, the DevOps team would be able to limit clones of files because the templating would maximize the reusability of the required core files. This was most evident with Terraform, which can natively leverage variable files to alter the behavior of the execution process. An engineer that needed to create or alter any cloud infrastructure would be able to submit a request with an approved set of variables to the execution process through a variable file. Then the delivery pipeline would pull the Terraform configuration files from the cloud infrastructure team's Git repository and the variable file from a completely separate Git repository, put them together, and run the required commands as a result. This would reduce the amount of time associated with a request, an approval, the Terraform configuration creation, and any other steps required by the cloud infrastructure team.

Another major benefit is that some of the tools and platforms either offered their own variable file process or allowed for variables to be leveraged in the core set of files. This would reduce the potential building requirements for the DevOps team to mitigate any lack of templating. The only thing that they would need to consider is when a tool or platform does not have templating built in, such as their legacy environment.

The ability to leverage declarative files for an execution process carries an inherent benefit of repeatability and reliability. The core execution engine is designed to consistently run the same process in the same way, with some behavior being altered based on values provided. Any team that is administrating a GitOps practice will have to balance the opinions implemented by the administration team and the configuration capability that is granted to the engineering teams. Some groups might decide that they will provide a single approved path that only allows for a change in the artifact name and version. Other groups might want the engineers to maintain control over every part of a deployment pipeline. But most groups will want a middle-ground approach, where they implement an approved path that allows the engineers to alter the behavior of their application but not the core functionality of the pipeline. The main issue with these different approaches, as with any declarative file approach, is the scalability of the operation.

If every step in a pipeline configuration file or variable file is statically defined, then any small iteration in either file will require its own version of a file. This is not too much of an issue if there is only one pipeline and once service moving through it. However, if there are multiple pipelines or multiple services being supported, then the potential for file sprawl will increase significantly. In order to make a repeatable and reliable process scalable, the use of templating must be implemented.

Templating is possible with declarative files, but it is not a native capability. Rather, a separate templating engine is required depending on which language is being used. One potential point of contention is which templating engine should be used, since some templating engines are provided on their own, and other templating engines are built into a platform or process that doesn't exist outside of the tool.

For example, in the case of Helm with Kubernetes, the templating of the YAML files is only related to the use of Helm.

Example 5.1: YAML resource file with basic go-templating:

```
# POD YAML
apiVersion: v1
kind: Pod
metadata:
  name: {{ .Values.name }}
spec:
  containers:
  - name: {{ .Values.name }}
    image: {{ .Values.image }}
```

Example 5.2: YAML file to declare variables passed to the resource file previously shown:

```
# VALUES FILE
name: app
image: library/nginx:latest
```

Example 5.1 and *5.2* are common with Helm, where there is a core resource file, a values file, and a variable convention to connect the two. The characters { { and } } are the indicators that the Helm templating engine looks for to override with a declared. Once the Helm templating engine overrides all required variables, it will then output a static manifest file that can be deployed to Kubernetes. Helm also provides some significant functionality for easier templating and scaling.

Example 5.3: YAML resource file showing advanced go-templating functions:

```
# POD YAML
apiVersion: v1
kind: Pod
metadata:
  name: {{ .Values.name }}
```

```
  namespace: {{ .Values.namespace | default "default" | lower
}}
spec:
  containers:
  - name: {{ .Values.name }}
    image: {{ .Values.image }}
    resources:
{{ .Values.resources | toYaml | indent 4 }}
```

Example 5.4: YAML file to declare variables passed to the resource file previously shown:

```
# VALUES FILE
name: app
image: library/nginx:latest
resources:
  limits:
    memory: "128Mi"
    cpu: "500m"
  requests:
    memory: "128Mi"
    cpu: "500m"
```

Example 5.4 shows how the values file remains the same as before, but the core resource file in *Example 5.3* now includes a few functions. The first function is the inline namespace reference, which tells Helm to pull the namespace from the values file, use `default` in the case that there is nothing in the namespace variable in the values file, and then convert every letter to a lowercase letter. Another function in *Example 5.3* is located in the `resources` section of the main resource file. This function allows the administrator to indicate that there is any number of objects that will exist in the map in the values file, and all objects in that list should be brought over, and they should all have the same level of indentation. This allows the Helm administrator to accommodate a dynamic list of properties that the user might provide in this section of the values file. These functions, as well as the plethora of other functions available with Helm, allow the Helm administrator to scale a Helm chart across multiple teams, services, and scenarios.

The main issue with Helm templating is that the templates only work in relation to Helm usage. If a user wants to define a pipeline through YAML files, they will have a difficult time leveraging the Helm templating engine to accomplish the desired outcome. This means that the DevOps team would have to find a different way to convert a YAML-defined pipeline configuration file into a template that can take in variables, which allows for scalability.

With JSON, there are some tools or platforms that allow for templating and variables, but those are limited to the execution engine of those tools, similar to the Helm use case shown previously. AWS CloudFormation and Terraform are both examples of tools that allow the use of JSON and variables, but they do not allow templating on generic JSON files that could be used for pipeline configuration code. In this case, the DevOps team would need to use an external tool that allows templating functionality on JSON files, such as Jsonnet.

Jsonnet is an open source project from Google that uses a templating engine that allows functions to be added into the JSON template file and outputs a static JSON file for use, similarly to how Helm uses a set of template and values files, runs them through a templating engine, and then outputs a static manifest YAML file for use. If the desired declarative language is JSON, then the only way to scale the potential JSON files is through a templating engine such as Jsonnet.

Although Jsonnet can output more than just JSON files, the idea remains the same, where Jsonnet uses a `.jsonnet` file, which contains code, to output desired static files.

Example 5.5: Jsonnet file:

```
{
  car1: {
    make: "Honda",
    model: "Civic",
    message: "My car is a " + self.make + " " + self.model
  },
  car2: self.car1 { make: "Toyota", model: "Corolla" }
}
```

Example 5.6: Jsonnet output file:

```
{
  "car1": {
    "make": "Honda",
    "model": "Civic",
```

```
      "message": "My car is a Honda Civic"
    },
  "car2": {
    "make": "Toyota",
    "model": "Corolla",
    "message": "My car is a Toyota Corolla"
    }
  }
```

Examples 5.5 and *5.6* show a basic set of files, the first being the Jsonnet file and the second being the output JSON file. The Jsonnet file starts with a function named Car that has two parameters named make and model. The function allows a user to specify the two parameters when calling the function, which is shown in the car1: Car('Honda', 'Civic') line. The output of the Jsonnet templating process creates a JSON file with the desired outcome of three car objects.

To allow a scalable set of JSON files, Jsonnet has the ability to use file inheritance, similar to modules or libraries for code.

Example 5.7: Jsonnet imports file:

```
local superhero = import 'superhero.libsonnet';


{
  'Batman': superhero['Batman'],
  'Spiderman': {
    attributes: [
      { outfit: 'suit', color: 'blue and red' },
      { vehicle: 'motorcycle', color: 'blue and red' }
    ],
    sidekick: 'no',
    city: 'queens'
  }
}
```

Example 5.8: Jsonnet libsonnet file:

```
// SUPERHERO.LIBSONNET
{
  'Batman': {
```

```
    attributes: [
        { outfit: 'suit and cape', color: 'black and yellow' },
        { vehicle: 'batmobile', color: 'black' }
    ],
    sidekick: 'yes',
    city: 'gotham'
},
'Superman': {
    attributes: [
        { outfit: 'suit and cape', color: 'blue and red' },
        { vehicle: 'none', color: 'none' }
    ],
    sidekick: 'yes',
    city: 'metropolis'
    }
}
```

Example 5.9: Jsonnet outputs file:

```
// OUTPUT.JSON FILE
{
  "Batman": {
    "attributes": [
      {
        "color": "black and yellow",
        "outfit": "suit and cape"
      },
      {
        "color": "black",
        "vehicle": "batmobile"
      }
    ],
    "city": "gotham",
    "sidekick": "yes"
  },
  "Spiderman": {
    "attributes": [
```

```
    {
        "color": "blue and red",
        "outfit": "suit"
    },
    {
        "color": "blue and red",
        "vehicle": "motorcycle"
    }
    ],
    "city": "queens",
    "sidekick": "no"
  }
}
```

Examples 5.7 through *5.9* show the use of a main Jsonnet file, a libsonnet file, and the output file. The libsonnet file contains objects in Jsonnet format, which is referenced in the Jsonnet file, shown in the first line as `local superhero = import 'superhero.libsonnet';`. This one line indicates that a file will be provided and information from that file can be used in the main Jsonnet file, which will then output the desired JSON file for later use. In the example, there is a libsonnet file that has the superheroes `Batman` and `Superman` defined. In the `superhero.jsonnet` file, there is another superhero defined, but there is a reference to `Batman` from the libsonnet file. The output shows that only `Batman` and `Spiderman` are shown in the output JSON file. To translate this outcome to pipeline code, the administrator might have a hard requirement on a stage that must exist in the pipeline, but the other stages can be provided by an engineer. The engineer would fill out information in the libsonnet file. Then the administrator would statically define the required stage in the Jsonnet file and reference the predefined stage in the libsonnet file. This will output the desired output JSON file that the pipeline would then use. Because of the use of the libsonnet file, the administrator can allow significant configurability for the engineer while also enforcing a standard path that must be followed.

The execution engine that consumes the output JSON file would be able to leverage the declarative nature of the JSON file to provide repeatability and reliability. The Jsonnet templating capabilities add scalability to the entirety of the process, especially with regard to GitOps. The libsonnet file update can trigger the templating process, with the output JSON file being automatically uploaded to a predefined Git repository. Then, when the output JSON file is uploaded to a Git repository, that can trigger the desired pipeline to deploy.

Summary

This chapter provided a deeper dive into some of the basics of building a declarative file. Templating was also introduced to show how to add scalability to the benefits of a declarative approach to delivery and deployments. Templating is accomplished through templating engines, which are directed at specific declarative languages, namely Jsonnet for JSON and Helm for YAML. You should now know the difference between a flat file and a nested file, an object and an array, and XML, JSON, and YAML file design and syntax.

The next chapter will go through the basics of originalist GitOps with Visual Studio Code, Minikube, ArgoCD, GitHub Actions, and Flux.

9
Originalist Gitops in Practice – Continuous Deployment

Previous chapters have focused on the hypothetical practices of GitOps and some of the questions and concerns that must be considered when wanting to adopt GitOps into a delivery or deployment process. The last chapter showed the practical portion of building out and templating declarative language files. Another major benefit of the previous chapter was showing a small portion of what is required when building out a process that heavily relies on declarative files.

This chapter will explore the originalist GitOps practice with some local development. The desired outcome of this chapter is to give you a better understanding of the fundamentals related to setting up originalist GitOps tools and processes.

In this chapter, we're going to cover the following main topics:

- Setting up minikube

- Setting up VSCode

- Setting up Kubernetes and Helm

- Local continuous deployment with GitOps file building

- Originalist GitOps with Argo CD

Setting up minikube

While the main DevOps team was working on the enablement process around declarative file creation for the rest of engineering, a small group from the DevOps team was working on the actual implementation process. The first of many issues that the DevOps team had was that the scope of required support and the available tools did not align. But once the team had looked into Ansible and Harness, although they both could support the required platforms and functionality, engineering leadership would require evidence of evaluation for the other tools available.

The team could not show a list of desired functionalities and a list of current functionalities and then assume that was sufficient. Especially since Ansible and Harness had associated costs, whether that was support or licensing cost, the team had to show due diligence in qualifying a tool or solution.

This meant that the DevOps team needed to present the tools that they evaluated, the functionality that was available, where the tools fell short, and why the tools would not work for the desired outcome. And because the GitOps tools that were first explored were Kubernetes-based, the team would show a few of the leading-option tools in that area. Argo CD is one of the most widely used GitOps tools for Kubernetes-only applications. To run this evaluation locally, the team would need to set up a Kubernetes cluster on their local computer, set up a code editor that could sync with a GitHub repository, create a basic Helm chart, and then enable Argo CD to deploy to the cluster.

Although originalist GitOps is a practice that has been previously covered in this book, it is one of the more common GitOps practices in use today. The main reason for this is a combination of the ease of deployment that is allowed by an operator-based deployment model and also because there is no licensing cost on true open source software. The moment that Kubernetes is implemented, most engineering teams look for a free and easy-to-use tool that automates part or all of a process. In the world of originalist GitOps, the automatable part is running the `kubectl apply` or `helm install/upgrade` command, and the process is the deployment. One of the most popular ways to test any of these tools for automation, specifically in the world of Kubernetes, is to run a local version of a Kubernetes cluster via the minikube solution.

Kubernetes clusters can become very expensive to run, especially when considering scaling, the lack of tuning and optimization, and all of the layers of abstraction that can hide resource misuse issues. As such, it is not recommended to run testing on a cost-based Kubernetes solution such as the ones offered by the different cloud providers. minikube allows an engineering team or Kubernetes administrator to test manifest types, Kubernetes support tools, scripts, and so on.

Setting up minikube on a computer is a fairly simple process that requires installing the following tools, in order.

> **Important note**
> All steps are based on using an Apple MacBook. Please research specific instructions if the operating system and device differ.

Homebrew

Homebrew is a popular package manager for Apple computers.

1. To download Homebrew, go to `https://brew.sh/` and follow the instructions to download and install:

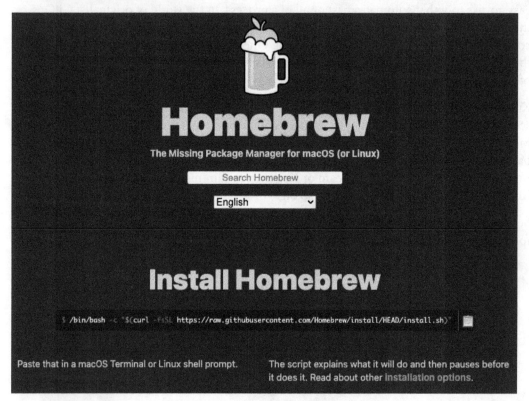

Figure 9.1 – Homebrew home page

This can also be installed via Homebrew by running the following:

```
/bin/bash -c "$(curl -fsSL https://raw.githubusercontent.com/
Homebrew/install/HEAD/install.sh)"
```

> **Important note**
>
> A Linux system will often use either `apt-get` or `yum` for its package managers. Windows users can leverage something such as `chocolatey`.

2. Installing Homebrew leads to a verbose output showing all of the actions
    being taken by the installer to both download and install Homebrew, as well as
    configuring it for use:

```
==> Tapping homebrew/core
remote: Enumerating objects: 129, done.
remote: Counting objects: 100% (129/129), done.
remote: Compressing objects: 100% (86/86), done.
remote: Total 893116 (delta 78), reused 76 (delta 43), pack-reused 892987
Receiving objects: 100% (893116/893116), 357.62 MiB | 481.00 KiB/s, done.
Resolving deltas: 100% (605918/605918), done.
From https://github.com/Homebrew/homebrew-core
 * [new branch]      master      -> origin/master
Checking out files: 100% (5749/5749), done.
HEAD is now at d5c0a4f9e6 gomplate: update 3.9.0 bottle.
==> Installation successful!

==> Homebrew has enabled anonymous aggregate formulae and cask analytics.
Read the analytics documentation (and how to opt-out) here:
  https://docs.brew.sh/Analytics
No analytics data has been sent yet (or will be during this `install` run).

==> Homebrew is run entirely by unpaid volunteers. Please consider donating:
  https://github.com/Homebrew/brew#donations

==> Next steps:
- Run `brew help` to get started
- Further documentation:
    https://docs.brew.sh
(base) →  ~ 
```

Figure 9.2 – Homebrew install output

3. Anytime a binary/CLI is downloaded and installed, an easy way to check that the
    command is in your path is by running a version check (such as `brew -v`):

```
(base) →  ~ brew -v
Homebrew 2.7.5
Homebrew/homebrew-core (git revision d5c0a4; last commit 2021-01-26)
```

Figure 9.3 – Homebrew command check

Now that we have successfully installed Homebrew, let's install our next tool: VirtualBox.

VirtualBox

Kubernetes is a container orchestrator that runs on top of actual servers, whether bare-metal or virtual. As such, a hypervisor software such as VirtualBox will be required. Homebrew has a formula available to install VirtualBox on your machine (`brew install --cask virtualbox`):

1. The website `https://formulae.brew.sh/` is where to search for available Homebrew formulas:

Figure 9.4 – The Homebrew Formulae page

2. Searching for any Homebrew formula is simple. Simply add the name of the desired technology into the search box, such as `VirtualBox`, and you will receive the required `brew` command:

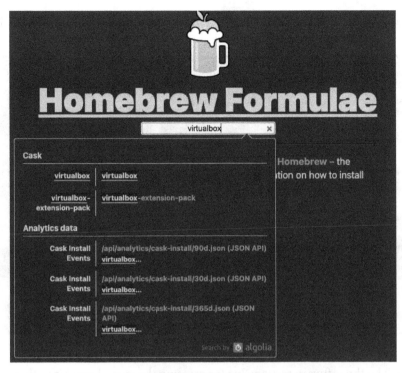

Figure 9.5 – VirtualBox Homebrew formula search

3. Every Homebrew formula has a details page that includes the brew command, the maintainers of the formula, the Git repository location, and so on:

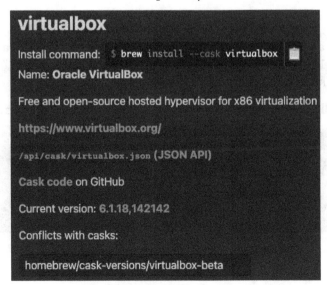

Figure 9.6 – VirtualBox formula page

4. Installing VirtualBox via Homebrew is done by running the `brew install --cask virtualbox` command:

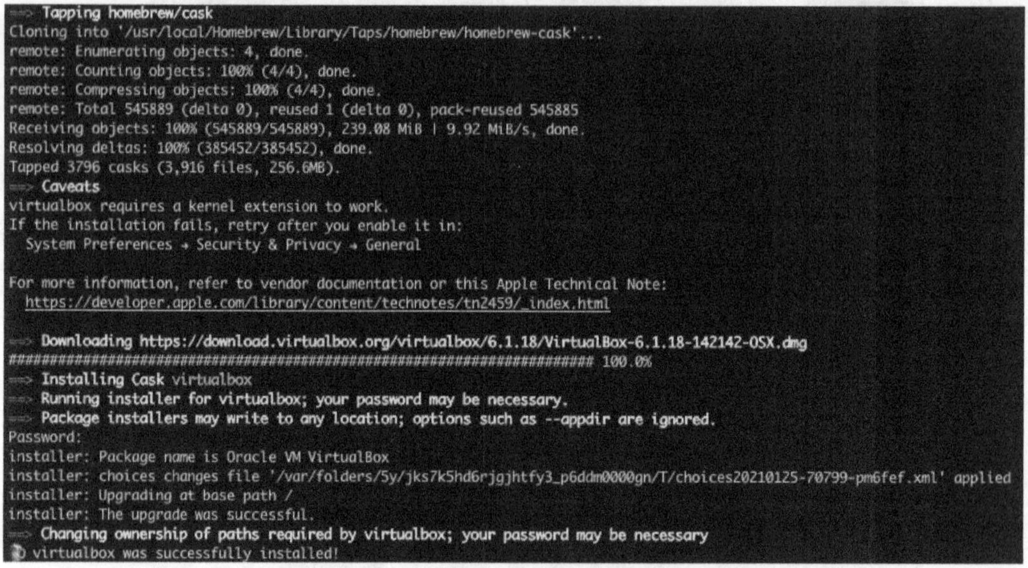

```
==> Tapping homebrew/cask
Cloning into '/usr/local/Homebrew/Library/Taps/homebrew/homebrew-cask'...
remote: Enumerating objects: 4, done.
remote: Counting objects: 100% (4/4), done.
remote: Compressing objects: 100% (4/4), done.
remote: Total 545889 (delta 0), reused 1 (delta 0), pack-reused 545885
Receiving objects: 100% (545889/545889), 239.08 MiB | 9.92 MiB/s, done.
Resolving deltas: 100% (385452/385452), done.
Tapped 3796 casks (3,916 files, 256.6MB).
==> Caveats
virtualbox requires a kernel extension to work.
If the installation fails, retry after you enable it in:
  System Preferences → Security & Privacy → General

For more information, refer to vendor documentation or this Apple Technical Note:
  https://developer.apple.com/library/content/technotes/tn2459/_index.html

==> Downloading https://download.virtualbox.org/virtualbox/6.1.18/VirtualBox-6.1.18-142142-OSX.dmg
################################################################ 100.0%
==> Installing Cask virtualbox
==> Running installer for virtualbox; your password may be necessary.
==> Package installers may write to any location; options such as --appdir are ignored.
Password:
installer: Package name is Oracle VM VirtualBox
installer: choices changes file '/var/folders/5y/jks7k5hd6rjgjhtfy3_p6ddm0000gn/T/choices20210125-70799-pm6fef.xml' applied
installer: Upgrading at base path /
installer: The upgrade was successful.
==> Changing ownership of paths required by virtualbox; your password may be necessary
🍺 virtualbox was successfully installed!
```

Figure 9.7 – VirtualBox formula install output

VirtualBox will be the software responsible for creating the virtual machines that Kubernetes will run on. Another tool required for the proper operation of Kubernetes and minikube is the Kubernetes command line.

kubectl

The main way that a user will interact with a Kubernetes cluster is through a CLI tool known as `kubectl` (`kubectl` can be pronounced as either "kube control," "kube-cuttle," "kube-cuddle," or "kube-ctl").

It can also be installed via Homebrew by running `brew install kubectl`:

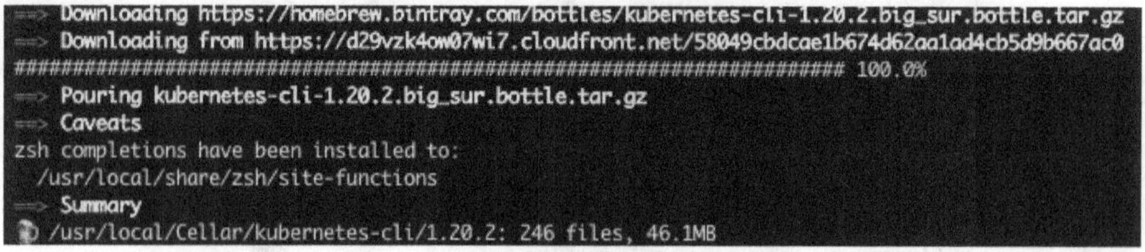

```
==> Downloading https://homebrew.bintray.com/bottles/kubernetes-cli-1.20.2.big_sur.bottle.tar.gz
==> Downloading from https://d29vzk4ow07wi7.cloudfront.net/58049cbdcae1b674d62aa1ad4cb5d9b667ac0
################################################################ 100.0%
==> Pouring kubernetes-cli-1.20.2.big_sur.bottle.tar.gz
==> Caveats
zsh completions have been installed to:
  /usr/local/share/zsh/site-functions
==> Summary
🍺 /usr/local/Cellar/kubernetes-cli/1.20.2: 246 files, 46.1MB
```

Figure 9.8 – kubectl formulae install output

Since Kubernetes requires both servers or virtual machines and command-line tools, it is a requirement to have both of these items installed before installing minikube. minikube is a smaller version of Kubernetes that can be operated with smaller resource requirements, such as CPU and memory. It is a very common tool for testing and proving out different Kubernetes requirements without needing to create and configure Kubernetes at a larger scale, which might result in cloud costs.

minikube

Now that Homebrew, VirtualBox, and `kubectl` are installed, minikube can be installed via the same Homebrew method as the other tools (that is, with `brew install minikube`):

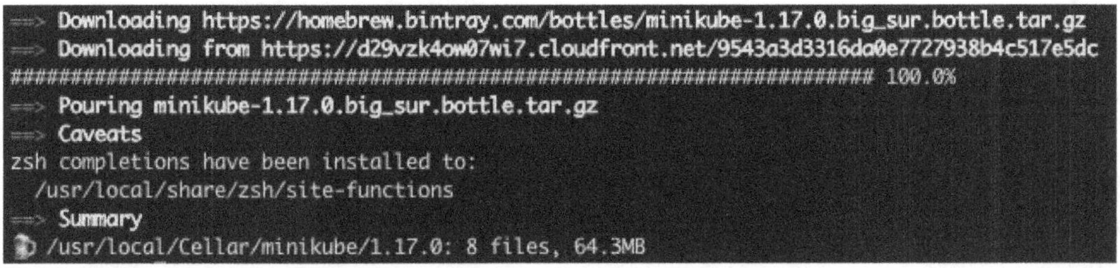

```
==> Downloading https://homebrew.bintray.com/bottles/minikube-1.17.0.big_sur.bottle.tar.gz
==> Downloading from https://d29vzk4ow07wi7.cloudfront.net/9543a3d3316da0e7727938b4c517e5dc
######################################################################## 100.0%
==> Pouring minikube-1.17.0.big_sur.bottle.tar.gz
==> Caveats
zsh completions have been installed to:
  /usr/local/share/zsh/site-functions
==> Summary
🍺 /usr/local/Cellar/minikube/1.17.0: 8 files, 64.3MB
```

Figure 9.9 – minikube formula install output

Once Homebrew and VirtualBox have been installed, minikube can be started. However, before starting minikube, some configuration changes might be required. For example, setting the memory limits for minikube to use can be accomplished through `minikube config set memory 8128`. Another configuration change that can be helpful is `minikube config set driver virtualbox`. After these settings are configured, running `minikube start` should bring up minikube with the updated settings:

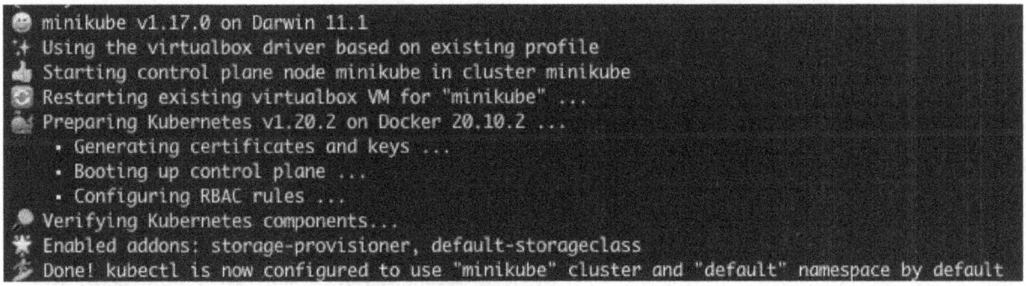

```
😄 minikube v1.17.0 on Darwin 11.1
✨ Using the virtualbox driver based on existing profile
👍 Starting control plane node minikube in cluster minikube
🔄 Restarting existing virtualbox VM for "minikube" ...
🐳 Preparing Kubernetes v1.20.2 on Docker 20.10.2 ...
    • Generating certificates and keys ...
    • Booting up control plane ...
    • Configuring RBAC rules ...
🔎 Verifying Kubernetes components...
🌟 Enabled addons: storage-provisioner, default-storageclass
🏄 Done! kubectl is now configured to use "minikube" cluster and "default" namespace by default
```

Figure 9.10 – minikube start output

Running `kubectl get ns` is an easy way to test that everything is set up correctly:

```
(base) → ~ kubectl get ns
NAME               STATUS    AGE
default            Active    43s
kube-node-lease    Active    44s
kube-public        Active    44s
kube-system        Active    44s
```

Figure 9.11 – kubectl command to list Kubernetes namespaces

The user can change the settings before they start minikube if they need to accommodate for machine resource limitations or requirements. But now that minikube is up and running, the next step in the process of GitOps is to get a code editor or **integrated development environment** (**IDE**) installed and configured.

Setting up VSCode

With minikube set up on their local computers, the team could now start to build out their Helm charts to deploy to the cluster. The goal would be to have a basic Helm chart, with some override requirements, that could be easily added to a Git repository, triggering Argo CD.

*The first thing that the team would need to do is get their local code editor set up for these processes. The majority of the team was using the open source **Visual Studio Code** (**VSCode**) that Microsoft released, which has a large marketplace of plugins. The team could leverage a Git repository plugin, a minikube and Kubernetes plugin, and a YAML and Helm plugin. With these plugins added to their VSCode, they would be able to validate the Helm charts, add visibility to the Git syncing process, and gain some version and resource visibility for the clusters.*

But to make sure that no issues were present in the process of building out the Helm charts, while also checking the basic GitOps process with Kubernetes, the team could use a plugin to execute the desired deployment process when a local file is saved. This would allow them to test their Helm charts and their GitOps process before moving everything into a Git repository and installing the GitOps tools into the cluster.

Having an extensive code editor is essential when building out declarative language files, especially in a predefined format such as Helm. Code editor plugins allow issues to be highlighted, validation on schema configuration, and easier troubleshooting requirements. A very popular and extensive code editor for developers is VSCode by Microsoft.

Before installing and configuring an originalist GitOps practice and toolset, local development and testing need to be done on the Helm chart. The main reason for this is that originalist GitOps tools only work on well-tested and already-developed Kubernetes manifests. All the building and testing of Helm charts must be accomplished outside of the GitOps tool and then added in for deployment automation.

Getting VSCode installed and configured on a computer is fairly simple.

Downloading and installing

Download VSCode from the `https://code.visualstudio.com/download` website, selecting the required package for your operating system:

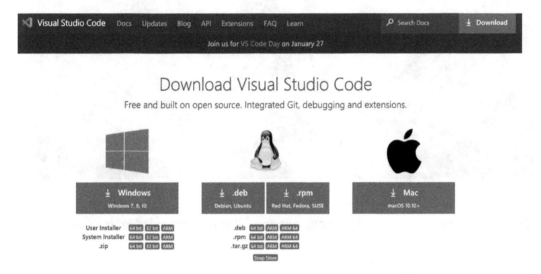

Figure 9.12 – VSCode download website

Once the installer is downloaded from Microsoft, follow the install steps with the default configuration:

Figure 9.13 – Downloaded ZIP file

After VSCode is installed on the machine, starting it will result in a welcome page for the user:

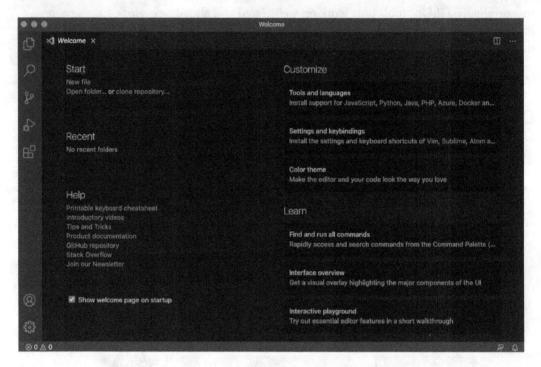

Figure 9.14 – VSCode welcome page

Once the installation is complete, we move on to adding recommended extensions.

Adding extensions

There are a few extensions that will help with the originalist GitOps process, such as Kubernetes, GitLens, YAML, and Helm. But there are other extensions and configurations that will be beneficial to add as well:

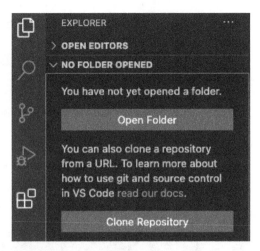

Figure 9.15 – VSCode file explorer

You can go to the VSCode extension explorer and type in the extensions you want to search for:

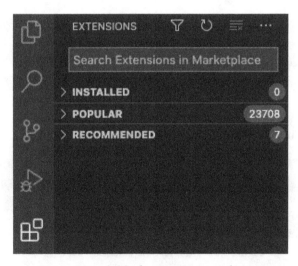

Figure 9.16 – VSCode extensions search page

The following are recommend extensions to install:

- **GitLens**: Provides powerful Git repository integrations and visibility:

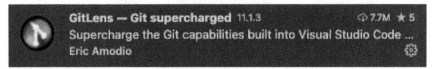

Figure 9.17 – GitLens extension page

- **Kubernetes**: A full toolset that has support for `kubectl`, Helm, and Docker:

Figure 9.18 – Kubernetes extension page

- **Helm Intellisense**: Adds VSCode IntelliSense for Helm charts and values files:

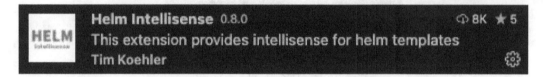

Figure 9.19 – Helm Intellisense extension page

- **indent-rainbow**: Makes indentation easier to read:

Figure 9.20 – indent-rainbow extension page

- **Run on Save**: Runs any command when a file is saved:

Figure 9.21 – Run on Save extension page

- **Bracket Pair Colorizer 2**: Matching brackets are identified with colors:

Figure 9.22 – Bracket Pair Colorizer extension page

- **Path Intellisense**: Autocompletes filenames:

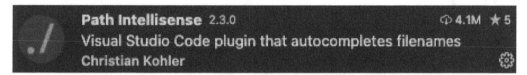

Figure 9.23 – Path Intellisense extension page

After VSCode is downloaded, installed, and configured, the next step will be to get a Helm chart ready for testing. This will require a folder that will eventually be used for syncing with a Git repository and triggering the originalist GitOps process.

Setting up Kubernetes and Helm

Although the other engineering teams had Helm charts already created, the DevOps team wanted to build out a core chart for easy templating. The goal would be to develop a core chart to test any business or engineering requirements in Kubernetes before passing them on to other teams. This would allow the enforcement of standards without burdening the engineering teams with additional configuration.

Using VSCode and minikube, the team could develop and deploy the Helm chart locally. Then, when they had a good base chart ready and deployable, they would initiate the sync with their Git repository to keep consistent track of any changes to the chart in the future. However, for the GitOps testing, they only needed a basic Helm chart to start with, allowing for some small overrides to show the functionality of the GitOps tool.

Since the feature sets of originalist GitOps tools are relatively limited in design and scope, an advanced Helm chart is not a requirement to test out functionality. But it is important to note that the time it takes to build and customize a Helm chart, regardless of how advanced the chart is, is not something that an originalist GitOps tool helps with. Therefore, any Helm chart that is built needs to be created locally and tested extensively before it can be connected to an originalist GitOps tool.

Creating a Helm chart is easy, but customizing it can be fairly daunting, depending on the desired behavior of the deployment.

Installing Helm

The first requirement is to install Helm. Similar to `kubectl`, VirtualBox, and `minikube`, Helm can be installed through the `brew install helm` command:

```
==> Pouring helm-3.5.0.big_sur.bottle.tar.gz
==> Caveats
zsh completions have been installed to:
   /usr/local/share/zsh/site-functions
==> Summary
🍺 /usr/local/Cellar/helm/3.5.0: 57 files, 43MB
```

Figure 9.24 – Helm formula install output

Creating a Helm chart

The `brew` install process for Helm will install the Helm CLI, which can be run with the `helm` command. To get started with Helm, the first thing to do is decide where you will create and store the Helm chart. A good location would be on your local desktop, which can be accessed by running the `cd ~/Desktop` command on a MacBook. Next, you will want to create a folder that can be used for connecting to a Git repository:

```
$ mkdir orig-gitops
$ cd orig-gitops
```

Figure 9.25 – Creating a directory for use

The command to create a basic Helm chart is to run `helm create APP_NAME` (which is `core-template` in the following example, created by running `helm create core-template`). Once that chart is created, you can open it in VSCode by opening VSCode and selecting **Open** from the **File** menu at the top of the screen and selecting the chart folder on your desktop. Once the folder is open in VSCode, you will be able to see the folder and file structure:

Figure 9.26 – VSCode file explorer with the Helm chart

Removing unwanted files

There are some files that exist in this folder that may not be needed. Examine the files before removing them to make sure that no required files are removed by accident. Deleting these files can be done by right-clicking the file and deleting it:

Figure 9.27 – The tests folder in a Helm chart

On clicking the file, a dialogue box will appear:

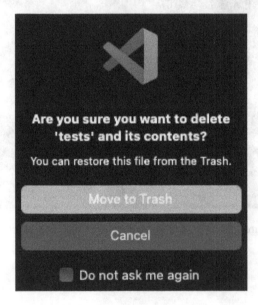

Figure 9.28 – The tests folder being deleted

Adding required files

A basic Helm chart will have a core set of files:

- **Template files**: Deployment, **HPA (Horizontal Pod Autoscaler)**, Ingress, Service
- **Parent files**: Chart, Values
- **Potentially useful files**: ConfigMap, Secret:

Figure 9.29 – Added files

Adding required information

The Kubernetes plugin for VSCode that was added earlier allows the quick creation of common Kubernetes resources. For this chart, we will need to create two files; one is `configmap.yaml` and the other is `secret.yaml`. Right-clicking on the **templates** folder and selecting **New File** is a quick way to accomplish this. Once the files are created, clicking on them will open them in the VSCode editor. You will need to change the file language to YAML, if required:

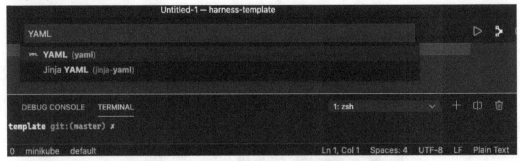

Figure 9.30 – Changing the file type

Now, when you type in the Kubernetes resource type, such as `configmap`, the VSCode YAML extension will show a list of options. When the `configmap` option is selected, the extensions will fill in the required information:

Figure 9.31 – ConfigMap auto-create

Once the Kubernetes ConfigMap is created, a Kubernetes Secret can be created in the same way:

Figure 9.32 – Secret auto-create

Exploring the files

Now that the appropriate files exist, it is important to explore the files in the chart to understand how they are built, especially with regard to Helm.

`deployment.yaml` is one of the main files to explore, since it is the file that defines how the actual application service is deployed to the cluster:

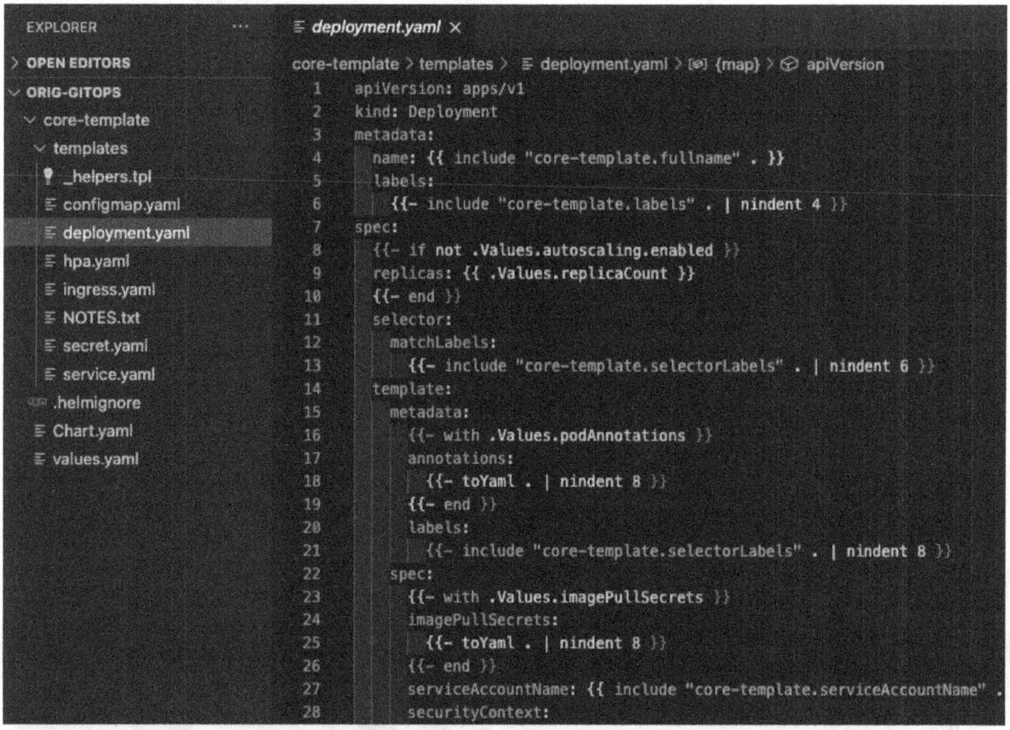

Figure 9.33 – The deployment.yaml file

A **Helm file** is a combination of a Kubernetes-specific structure and static information with some Helm and Go templating built in. Any time there is a `{{ }}` set in the Kubernetes file, that is where Helm or Go templating is involved.

There are some parts of the deployment file that are not needed, such as the node selector, affinity, toleration, security context, and service account. Although there are good use cases for these items, in a basic setup they are not needed:

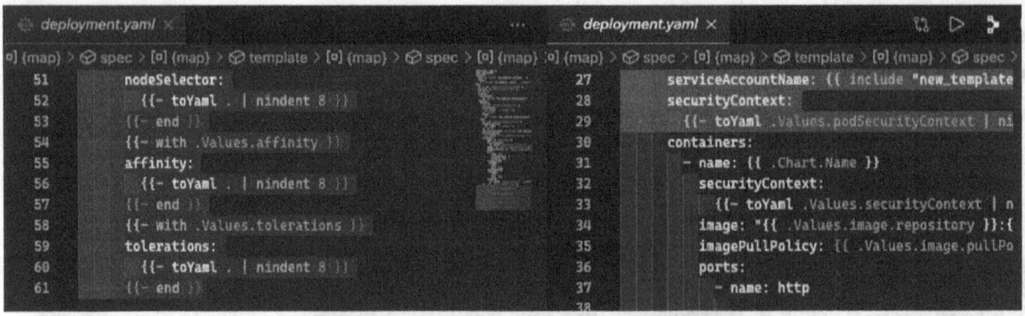

Figure 9.34 – Removal of extra sections

Explore the rest of the files and remove any unnecessary fields from the files, if needed.

Testing the deployment

Once the files have been created and configured, deploy out the Helm chart through the following steps:

1. Make sure that minikube is started:

```
minikube v1.17.0 on Darwin 11.1
Using the virtualbox driver based on existing profile
Starting control plane node minikube in cluster minikube
Updating the running virtualbox "minikube" VM ...
Preparing Kubernetes v1.20.2 on Docker 20.10.2 ...
Verifying Kubernetes components...
Enabled addons: storage-provisioner, default-storageclass
Done! kubectl is now configured to use "minikube" cluster and "default" namespace by default
```

Figure 9.35 – minikube start output

2. Then, you will need to make sure that you are in the folder that your Helm chart is in by running the `cd ~/Desktop/CHART_NAME` command.
3. Next, you will install the Helm chart into the Kubernetes cluster using `helm install . -generate-name`:

```
NAME: chart-1611811453
LAST DEPLOYED: Wed Jan 27 23:24:15 2021
NAMESPACE: default
STATUS: deployed
REVISION: 1
TEST SUITE: None
```

Figure 9.36 – Helm install output

4. Check that everything was created in the cluster using `kubectl get all`:

```
NAME                                           TYPE        CLUSTER-IP       EXTERNAL-IP   PORT(S)   AGE
service/chart-1611811453-core-template         ClusterIP   10.100.113.255   <none>        80/TCP    4m
service/kubernetes                             ClusterIP   10.96.0.1        <none>        443/TCP   2d1h

NAME                                                  READY   UP-TO-DATE   AVAILABLE   AGE
deployment.apps/chart-1611811453-core-template        0/1     0            0           4m

NAME                                                     DESIRED   CURRENT   READY   AGE
replicaset.apps/chart-1611811453-core-template-7b94d774b4   1         0         0       4m
```

Figure 9.37 – kubectl search command output

With the basic chart being created and deployed, the next step is to get a local version of originalist GitOps implemented to show how changes will affect deployments and what guardrails need to be set up to avoid issues with deployments or the cluster.

Local continuous deployment with GitOps file building

Setting up a local version of continuous deployment to the minikube cluster just requires some file change event to trigger the deployment. The DevOps team needed to figure out a way to look for those changes, which would mainly happen based on a file save. Since VSCode is very extensible, there is already a community extension that allows for a file save to trigger a designated deployment.

After installing the extension in VSCode, the DevOps team would then need to configure the appropriate file set to trigger the deployment, and then they would have to have the extension run the designated Helm command. This process would shave a few seconds off of each round of testing, since the team wouldn't have to switch to their terminal to execute the Helm command for the deployment.

Once the local version of the Helm chart was consistently healthy for every deployment, then the team would be able to add it to the desired Git repository. That would allow the desired tooling to be installed into the minikube cluster to demonstrate the GitOps process.

The ability to trigger an action based on a file save can be helpful for many reasons. In the case of testing a Helm chart, the process can save time in the deployment execution step. And when a Git repository is included in the process, the desired Git command could be automatically run upon saving to sync the files to Git.

VSCode has a few extensions that can help with this, one of which was installed previously, called **Run on Save**. This extension accomplishes exactly what the title says, and is easy to configure.

Starting minikube

If any testing is being done with minikube, always check to make sure that it is running first:

```
$ minikube start
```

```
minikube v1.17.0 on Darwin 11.1
Using the virtualbox driver based on existing profile
Starting control plane node minikube in cluster minikube
minikube 1.17.1 is available! Download it: https://github.com/kubernetes/minikube/releases/tag/v1.17.1
To disable this notice, run: 'minikube config set WantUpdateNotification false'

Updating the running virtualbox "minikube" VM ...
Preparing Kubernetes v1.20.2 on Docker 20.10.2 ...
Verifying Kubernetes components...
Enabled addons: storage-provisioner, default-storageclass
Done! kubectl is now configured to use "minikube" cluster and "default" namespace by default
```

Figure 9.38 – minikube start output

Run on Save configuration

To test out the **Run on Save** functionality, the extension needs to be properly configured:

1. In VSCode, there is an extensions icon on the left navigation bar. Once that is selected, there will be a list of installed extensions. From this list, select the **Run on Save** extension:

Figure 9.39 – VSCode extensions list

2. Next to the two blue buttons titled **Disable** and **Uninstall**, there is a gear icon to access the extension settings. Click on it:

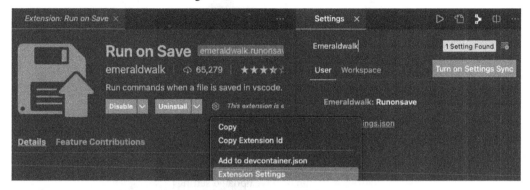

Figure 9.40 – Run on Save settings list

3. The settings for the extension will show up as a `settings.json` file, which is where the command to run the designated command will live:

Figure 9.41 – Run on Save settings page

4. To run the command, a `commands` key will need to be added with a JSON object nested inside of an array. The `match` key will indicate which file saves run the commands and the `cmd` key is the command that is run. The command to test this out will be `helm upgrade`, which requires the path to the chart and the name of the release. To get the name of the release, you will need to run `helm list` in your terminal, which will show the releases in the cluster:

```
NAME                NAMESPACE    REVISION    UPDATED                           STATUS      CHART              APP VERSION
chart-1611811453    default      1           2021-01-27 23:24:15.132401 -0600 CST   deployed    core-template-0.1.0    1.16.0
```

Figure 9.42 – Chart release list

5. Adding the command to upgrade the release as the **Run on Save** extensions execution command will allow continuous deployments of the Helm chart:

```
{
    "window.zoomLevel": -1,
    "emeraldwalk.runonsave": {
        "commands": [
            {
                "match": ".+",
                "cmd": "helm upgrade chart-1611811453 /path/to/chart/orig-gitops/core-template"
            }
        ]
    }
}
```

Figure 9.43 – Run on Save updated settings

6. With this setting now set and saved, the test will require viewing the history of the release in the cluster, changing the chart file and saving it, and then checking for a new revision of the release using `helm history <RELEASE NAME>`:

REVISION	UPDATED	STATUS	CHART	APP VERSION	DESCRIPTION
1	Wed Jan 27 23:24:15 2021	deployed	core-template-0.1.0	1.16.0	Install complete

Figure 9.44 – Helm chart release history

7. Once the extension settings are saved, the changing and saving of a file will trigger a deployment:

```
17  serviceAccount:                                              17  serviceAccount:
18    # Specifies whether a service account should be            18    # Specifies whether a
19    create: true                                               19    create: false
20    # Annotations to add to the service account                20    # Annotations to add to
21    annotations: {}                                            21    annotations: {}
22    # The name of the service account to use.                  22    # The name of the service
23    # If not set and create is true, a name is gener           23    # If not set and create
24    name: ""                                                   24    name: ""
```

Figure 9.45 – Changing a variable

8. Once the file has been changed, saving the file will trigger a new deployment:

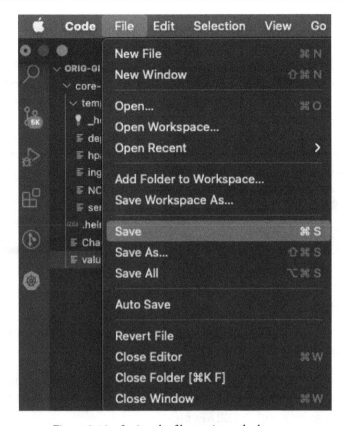

Figure 9.46 – Saving the file to trigger deployment

9. Checking the history of the release in Kubernetes will show the desired output:

REVISION	UPDATED	STATUS	CHART
1	Wed Jan 27 23:24:15 2021	deployed	core-template-0.1.0
2	Thu Jan 28 21:29:03 2021	failed	core-template-0.1.0

Figure 9.47 – Checking the Helm chart release history

This extension being set will allow rapid testing to be accomplished because every save will deploy a new version to the cluster. The match condition allows the user to specify which files should trigger a deployment if there are specific files that should be targeted. However, in the case of originalist GitOps, if any file is updated, then the execution is triggered.

The next step in the process is to configure the proper Git sync between the local chart and a repository for use with Argo CD. The process to get everything configured is fairly straightforward.

GitHub

GitHub allows both private and public Git repositories, which can easily be used for GitOps. Setting up a Git repository in GitHub requires just a set of simple clicks. You can start by going to www.github.com and creating an account if you don't have one already. When creating your first repository, make sure to give it a name that is easy to remember, and a good convention would be to keep the name in all-lowercase and use _ or - instead of spaces. For example, hello-world is better than Hello World:

Figure 9.48 – Creating a GitHub repository

GitHub access

With a Git repository now created, an SSH key will be required to connect the local files and also for the originalist GitOps tool:

1. Create an SSH key using the ssh-keygen -t rsa -m PEM command. When creating the key, the prompt will ask you where you want to save the file. A common location is /Users/YOUR_USER/.ssh:

```
Generating public/private rsa key pair.
Enter file in which to save the key (/Users/       /.ssh/id_rsa): /Users/        .ssh/rrs_book
Enter passphrase (empty for no passphrase):
Enter same passphrase again:
Your identification has been saved in /Users/       /.ssh/rrs_book.
Your public key has been saved in /Users/        /.ssh/rrs_book.pub.
The key fingerprint is:
SHA256:xv0GeJu0Q8FIgbUh+obFyZm7ZPosbEr6IGXVbBI1D3c
The key's randomart image is:
+----[RSA 3072]----+
|  .*+*= . E       |
| +.=.* * .        |
|. B + = .         |
| + o o o o        |
|. 0   . S +       |
| B . . . . =      |
|=..   .   o o     |
|=*    . o .       |
|=o+      . .      |
+----[SHA256]-----+
```

Figure 9.49 – SSH key creation output

2. In GitHub, under the user icon in the top right, there is a **Settings** option. Click on it:

Figure 9.50 – GitHub user settings

3. Once selected, there will be a list of menu options on the left navigation, with an option for **SSH and GPG keys**:

Figure 9.51 – GitHub SSH and GPG key location

This is the section that allows the user to add a new SSH key:

Figure 9.52 – Adding an SSH key

On MacBooks, there is a built-in function called pbcopy that adds the output of a command in the terminal to the clipboard, which can then be pasted in the new SSH key box in GitHub. It is important to note that the public key is required for this, meaning the key that ends in .pub. The `cat ~/.ssh/id_rsa.pub | pbcopy` command will copy the key to your clipboard so that you can paste it into GitHub easily:

SSH keys / Add new

Title

gitops-key

Key

ssh-rsa
fdhtwerybadsfgvsdytbsdhtbsdtybsdyrbsdfghbsdrtybsdthsbdfhw456w45yw45ybwetryw45bywertybwe56bwe
tybw45ytwbe56bwye56saw5eq234657nu46w8nwb4rer6qw34g6w457638568nert6gq34g6w4h67nw45e6b
q34b7w46e74w56nq347wetryuer6h73468n3568m956m8nw457bq46457n34568eetykirjt65y63452183745
9v817qwnetcy23f5n8712345c8n217435087n1243085v713485n7c13847c51032849dk50192k4750c1m8475
018724n50n172345vc7hZEIxAbN8GqGq8gw1DiZ4PA7EEmGmNc1onMI6GlOiA8WxGPZmr9LyY7g00wH/vsJ5f
pcdZxnS9+xV/H3CtD2hVF4GMTRUjTISj6ZzdfP925h8olzaobyHK1cfKpU9aKJf6CyQ7sB+3yqPfpJ25rqixbTOY8
Owc= user@laptop

Add SSH key

Figure 9.53 – SSH key box

4. After that is added and saved, the next step is to sync the local chart to the Git repository.

    In the directory where the chart lives, which has been ~/Desktop/CHART_NAME so far, run the `git init` Git initialization command. Once that is done, then the connection to the remote repository needs to be set up:

Figure 9.54 – GitHub code download and HTTPS link

5. The next step is to sync the GitHub repository with the files on the machine using the `git remote add origin <https/ssh path to repo>` command. After the connection is made, running the `git add .` command will stage the files to sync to the Git repository. Next, the staged files need to be committed to the Git repository with a commit message using `git commit -m "<relevant commit message>"`. Lastly, pushing the committed changes to the Git repository can be accomplished by running the `git push -u origin master` command. By storing the private SSH key that was created in the first step in the `~/.ssh` folder, the Git CLI will leverage that SSH key to authenticate with the Git repository. This is possible because the public version of your key was added to Git in a previous step:

```
Enumerating objects: 12, done.
Counting objects: 100% (12/12), done.
Delta compression using up to 8 threads
Compressing objects: 100% (12/12), done.
Writing objects: 100% (12/12), 4.62 KiB | 4.62 MiB/s, done.
Total 12 (delta 0), reused 0 (delta 0)
remote:
remote: Create a pull request for 'master' on GitHub by visiting:
remote:        https://github.com/bfeuling01/rrs-book/pull/new/master
remote:
To https://github.com/bfeuling01/rrs-book.git
 * [new branch]      master -> master
Branch 'master' set up to track remote branch 'master' from 'origin'.
```

Figure 9.55 – Output of Git commit and push

6. Once that is successful, the Git repository master branch should contain the same files as the Helm chart on the laptop. Changing the branch in the Git repository will show this:

Figure 9.56 – GitHub repository updated with files

With the Helm chart created, the Git repository set up for synchronization, and minikube being successfully deployed to, the GitOps tools can be added to the cluster to show the originalist GitOps process in action.

Originalist GitOps with Argo CD

After the Git repository, the minikube cluster, and the Helm chart were set up, the DevOps team could now start working on getting the GitOps tools implemented. The first tool to test out for the team was Argo CD. Although Argo CD wasn't the first GitOps tool available, it was definitely the most popular tool. The goal of implementing Argo CD into the cluster was to show both how easily and quickly the solution can be set up and working, and how much work is required to make sure that Argo CD fits into their company's requirements.

Argo CD is open source, which is typically sought after by engineering teams because it is free. There are only a few commands that need to be run to get Argo CD working in a cluster. But the DevOps team would need to document how they would secure Argo CD against unauthorized use, how to leverage multiple clusters, how to bring on multiple Helm charts, and how to deploy across all environments sequentially with approvals. After this testing and documentation were completed, the team could then move on to GitOps, which covers all of their technology stack.

The Argo project, which was built by Applatix back in 2017, was intended to be a container-native workflow engine. Many different tools have come out of the Argo project over the past few years, but the most popular has been Argo CD. Argo CD is Kubernetes operator-based, easy to install, and quick to configure.

Starting minikube

Any time that minkube is being used, always make sure that it is started before testing anything:

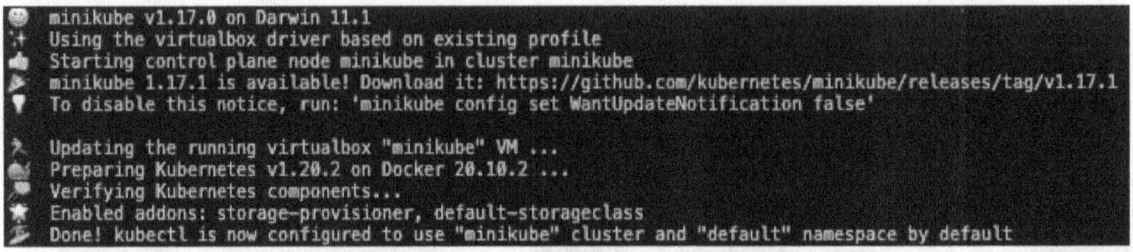

```
😊  minikube v1.17.0 on Darwin 11.1
✨  Using the virtualbox driver based on existing profile
👍  Starting control plane node minikube in cluster minikube
📦  minikube 1.17.1 is available! Download it: https://github.com/kubernetes/minikube/releases/tag/v1.17.1
💡  To disable this notice, run: 'minikube config set WantUpdateNotification false'

🔄  Updating the running virtualbox "minikube" VM ...
🐳  Preparing Kubernetes v1.20.2 on Docker 20.10.2 ...
🔎  Verifying Kubernetes components...
🌟  Enabled addons: storage-provisioner, default-storageclass
🏄  Done! kubectl is now configured to use "minikube" cluster and "default" namespace by default
```

Figure 9.57 – minikube startup output

Disabling the Run on Save extension

Since Argo CD is being used for the automated deployments, the Run on Save extension can be disabled:

1. In the VSCode extensions, select **Run on Save**:

Figure 9.58 – The Run on Save extension

2. Click the **Disable** button and then reload VSCode:

Figure 9.59 – The Run on Save extension is disabled

Installing Argo CD

Argo CD has an easy install process. The **Getting Started** page (`https://argoproj.github.io/argo-cd/getting_started/`) will give some basic instructions on how to get Argo CD installed into the cluster, such as running `kubectl create namespace argocd` to create the Argo CD namespace, and then running the install command:

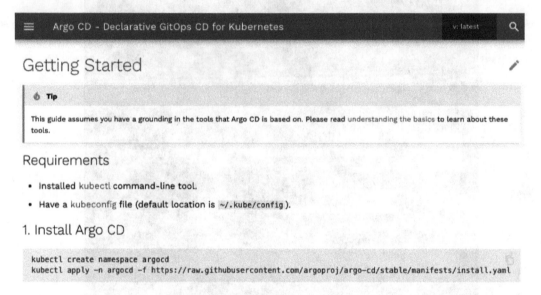

Figure 9.60 – Argo CD Getting Started page

Once the install command for Argo CD is run, the output will show all of the installed components in the Kubernetes cluster:

```
Warning: apiextensions.k8s.io/v1beta1 CustomResourceDefinition is deprecated in v1.
customresourcedefinition.apiextensions.k8s.io/applications.argoproj.io created
customresourcedefinition.apiextensions.k8s.io/appprojects.argoproj.io created
serviceaccount/argocd-application-controller created
serviceaccount/argocd-dex-server created
serviceaccount/argocd-redis created
serviceaccount/argocd-server created
role.rbac.authorization.k8s.io/argocd-application-controller created
role.rbac.authorization.k8s.io/argocd-dex-server created
role.rbac.authorization.k8s.io/argocd-redis created
role.rbac.authorization.k8s.io/argocd-server created
clusterrole.rbac.authorization.k8s.io/argocd-application-controller created
clusterrole.rbac.authorization.k8s.io/argocd-server created
rolebinding.rbac.authorization.k8s.io/argocd-application-controller created
rolebinding.rbac.authorization.k8s.io/argocd-dex-server created
rolebinding.rbac.authorization.k8s.io/argocd-redis created
rolebinding.rbac.authorization.k8s.io/argocd-server created
clusterrolebinding.rbac.authorization.k8s.io/argocd-application-controller created
clusterrolebinding.rbac.authorization.k8s.io/argocd-server created
configmap/argocd-cm created
configmap/argocd-gpg-keys-cm created
configmap/argocd-rbac-cm created
configmap/argocd-ssh-known-hosts-cm created
configmap/argocd-tls-certs-cm created
secret/argocd-secret created
service/argocd-dex-server created
service/argocd-metrics created
service/argocd-redis created
service/argocd-repo-server created
service/argocd-server created
service/argocd-server-metrics created
deployment.apps/argocd-dex-server created
deployment.apps/argocd-redis created
deployment.apps/argocd-repo-server created
deployment.apps/argocd-server created
statefulset.apps/argocd-application-controller created
```

Figure 9.61 – Argo CD install command output

Installing Argo CD CLI

After the Argo namespace and solution are installed in the cluster, Argo CD CLI will need to be installed to interact with portions of the Argo CD instance:

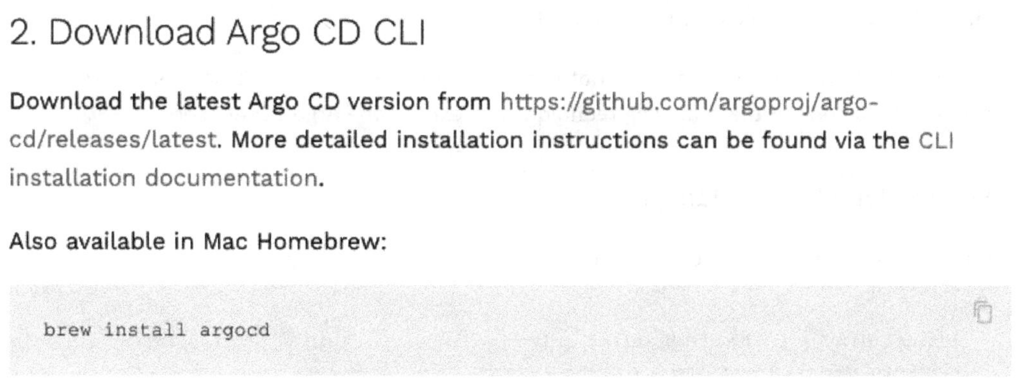

2. Download Argo CD CLI

Download the latest Argo CD version from https://github.com/argoproj/argo-cd/releases/latest. More detailed installation instructions can be found via the CLI installation documentation.

Also available in Mac Homebrew:

```
brew install argocd
```

Figure 9.62 – Argo CD CLI install page

The output from the Homebrew command will show which version of the Argo CD CLI is being installed:

```
==> Downloading https://homebrew.bintray.com/bottles/ar
==> Downloading from https://d29vzk4ow07wi7.cloudfront.
#########################################################
==> Pouring argocd-1.8.3.big_sur.bottle.tar.gz
==> Caveats
zsh completions have been installed to:
  /usr/local/share/zsh/site-functions
==> Summary
🍺 /usr/local/Cellar/argocd/1.8.3: 8 files, 69.7MB
```

Figure 9.63 – Homebrew install of Argo CD CLI

Running a quick Argo CD version check using the `argocd` version command will show that Argo CD CLI is installed correctly:

```
argocd: v1.8.3+0f9c684.dirty
  BuildDate: 2021-01-23T03:52:47Z
  GitCommit: 0f9c68427882bf4633d395cbfcd7c9271795fd9b
  GitTreeState: dirty
  GoVersion: go1.15.7
  Compiler: gc
  Platform: darwin/amd64
```

Figure 9.64 – Argo CD CLI version

Argo CD API server access

The Argo CD API server needs to be exposed to allow interactions with the API:

3. Access The Argo CD API Server

By default, the Argo CD API server is not exposed with an external IP. To access the API server, choose one of the following techniques to expose the Argo CD API server:

Service Type Load Balancer

Change the argocd-server service type to `LoadBalancer`:

```
kubectl patch svc argocd-server -n argocd -p '{"spec": {"type": "LoadBalancer"}}'
```

Figure 9.65 – Exposing the Argo CD API server

Logging in via the CLI

With the Argo CD API now exposed, the next step is to log in to Argo CD via the CLI and update the password:

> Argo CD 1.8 and earlier
>
> The initial password is autogenerated to be the pod name
> This can be retrieved with the command:

```
kubectl get pods -n argocd -l app.kubernetes.io/name=argo
```

Using the username `admin` and the password from above, hostname:

```
argocd login <ARGOCD_SERVER>  # e.g. localhost:8080
```

Change the password using the command:

```
argocd account update-password
```

Figure 9.66 – Argo CD CLI login and changing the password

Registering the cluster

To deploy to a designated cluster, that cluster must be registered with Argo CD. This is done by getting the desired context and passing that context to Argo CD CLI, which then installs an admin role service account:

5. Register A Cluster To Deploy Apps To (Optional)

This step registers a cluster's credentials to Argo CD, and is only necessary when deploying to an external cluster. When deploying internally (to the same cluster that Argo CD is running in), https://kubernetes.default.svc should be used as the application's K8s API server address.

First list all clusters contexts in your current kubeconfig:

```
kubectl config get-contexts -o name
```

Choose a context name from the list and supply it to `argocd cluster add CONTEXTNAME`.

Figure 9.67 – Registering a cluster with Argo CD

Registering the Kubernetes cluster with Argo CD will install the required service account, ClusterRole, and ClusterRoleBinding for Argo CD to work in that cluster. This is done by running `argocd cluster add`, which will ask for the context name of the cluster and give a list of contexts to use. Then, running the same `argocd cluster add` `CONTEXT_NAME` will install the required pieces for you:

```
INFO[0000] ServiceAccount "argocd-manager" created in namespace "kube-system"
INFO[0000] ClusterRole "argocd-manager-role" created
INFO[0000] ClusterRoleBinding "argocd-manager-role-binding" created
Handling connection for 8080
Cluster 'https://192.168.99.103:8443' added
```

Figure 9.68 – Cluster register output

Creating an application from a Git repository

The next step is to connect Argo CD to the Git repository and begin deploying. A default local instance of Argo CD should be available at `localhost:8080`. Then, the credentials from *Logging in via the CLI* section previously will be used to log in:

Figure 9.69 – The Argo CD login page

After logging in to the Argo CD instance, the first section to visit is the **Settings** page:

Figure 9.70 – The Argo CD Settings page

The **Settings** page allows the user to connect to Git repositories for access to Kubernetes manifests.

Figure 9.71 – The Argo CD Repositories page

When adding a new Git repository, the user will be prompted with how to connect to the Git repository:

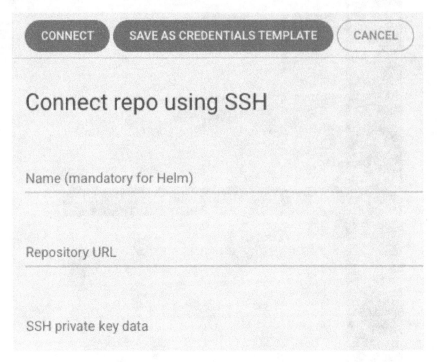

Figure 9.72 – Connecting to a Git repository

All Git repositories have a simple way of getting the connection information for the desired repository. The user will need to grab the required clone URL from the Git repository and add it to Argo CD:

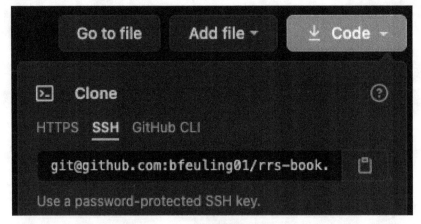

Figure 9.73 – Git repository connection URL

Once the Git repository URL is added to Argo, the user will need to add their private SSH key to allow Argo CD to monitor and pull the code files from the designated Git repository. This is the same `~/.ssh/id_rsa` file that was created for the GitHub repository connection process:

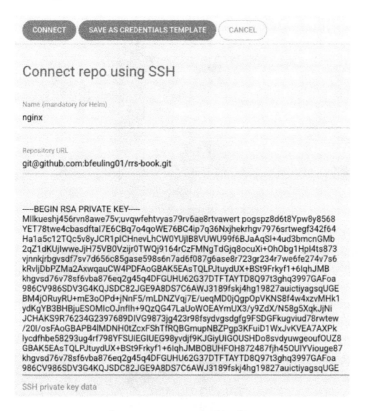

Figure 9.74 – Private RSA key

After the Git repository is connected, the user will be able to create an application in Argo CD using the Git repository. To do this, when the Git repository is connected, there is a three-dots settings menu at the bottom right of the repository in Argo CD that allows you to create the application. Once that is clicked, an application creation screen will appear that allows you to specify any required settings:

- **Application Name**: `general-application`.
- **Project**: default.
- **SYNC POLICY: Automatic**.
- **Prune Resources: True**.
- **Self Heal: True**.

- **Destination Cluster**: Set to Name instead of URL and select the first name from the dropdown.

- **Namespace**: default

- **Path**: . :

Figure 9.75 – Creating an app from a repository

Once the application settings are saved, you should be able to see the application being created in the main Argo CD dashboard. You can then click on the application to see a more expanded view of the application components:

Figure 9.76 – App deployment and health

Pushing a change

To see the GitOps process auto-deploy a new change to the Helm chart, make a change in the chart and push it to GitHub. In this example, the change is being made to the values.yaml file. Once pushed to GitHub, the change should trigger a sync in Argo CD:

```
# Declare variables to be passed into your templates.

replicaCount: 1

image:
  repository: nginx
  pullPolicy: Always       You, seconds ago • Uncommitted changes
  # Overrides the image tag whose default is the chart appVersion.
  tag: ""
```

Figure 9.77 – Changing a value in a file

Once a file in GitHub is changed and committed, Argo CD will see the updated file and trigger a deployment to sync the file to the cluster:

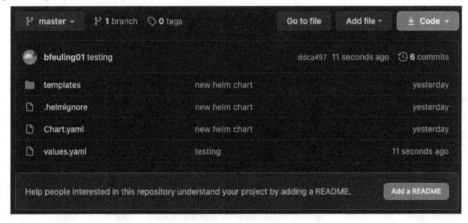

Figure 9.78 – Updated file in the Git repository

The user will be able to see the progress of the new install and the outcome once it is completed:

Figure 9.79 – New deployment triggered

Those are the basic requirements to get Argo CD set up inside of a cluster. For future Helm charts or Kubernetes manifests, steps in *Creating an application from a Git repository* and *Pushing a change* will need to be repeated for every application and manifest set.

Other than adding more manifest sets, the only other piece to consider is when it is needed to deploy to multiple clusters. One option is to install a different Argo CD set in each cluster and then leverage different logins for each cluster. That way can be rather tedious, but it would allow the same chart to be deployed to multiple clusters in parallel, if that was a desire. The alternative is to use a combination of an App of Apps deployment pattern and leveraging remote connection to each cluster's master node. An App of Apps is just that, an application in Argo CD that deploys other applications. The nesting of applications is a common way to overcome the lack of native multi-cluster and multi-environment support in Argo CD. The tutorial at `https://argoproj.github.io/argo-cd/operator-manual/declarative-setup/#app-of-apps` walks through this concept more.

Summary

A tool such as Argo CD has a narrow support scope, being limited only to Kubernetes. But this narrow scope allows Argo CD to offer a quicker time-to-value, from installing to deploying. The capabilities of the tool have rightly earned it a reputation as one of the elite open source originalist GitOps tools widely used today.

The next chapter will cover how to implement verified GitOps. Integrating with multiple clouds and platforms, supporting an array of application architectures, and enforcing security and compliance are essential in today's technology-forward world. Verified GitOps aims to support these requirements in a declarative and automated way.

10

Verified GitOps Setup – Continuous Delivery GitOps with Harness

When considering any kind of automation, understanding and mapping out the process is a hard requirement. Automation applied to an undefined process will result in gaps or holes that require manual intervention. Adopting a GitOps practice should start with documenting where automation is desired, how to define it in code, and how to trigger the execution.

The previous chapter walked through setting up originalist GitOps using Minikube, Helm, and ArgoCD. This chapter will look to set up verified GitOps, starting out with how to map the process and how to use declarative language for repeatability and reliability. Additionally, since an open source tool was used in the last chapter to show originalist GitOps, this chapter will use a vendor-based tool for verified GitOps.

In this chapter, we're going to cover the following main topics:

- Mapping out the process
- One manifest or many
- A manifest for integrations
- A manifest for configuration
- A manifest for execution
- A manifest for delivery
- Verified GitOps with Harness

Mapping out the process

The DevOps team had already settled on two different potential solutions for applying GitOps practices to their delivery process. One of the tools, Ansible, is open source and allows the team to achieve their GitOps requirements without any licensing cost. The other tool, Harness, is not an open source tool and requires the company to purchase licenses. Harness also allows the team to achieve continuous delivery with GitOps, and with significantly less setup, configuration, and administration than Ansible.

As the deliberation over whether Harness will be purchased or not continues, the DevOps team is finding that they have less and less time to get their first iteration of the process ready and usable by the other engineers. To make the solution setup and configuration easier, they decide to map out their process, tools, platforms, security requirements, and so on. They know that a cloud platform and their data centers will all need to be supported. The artifacts for the different endpoints are stored in a single artifact repository. Every application has two non-production environments and one production environment. Some of the platforms allow for automated scaling, such as Kubernetes and the serverless option, while the traditional application needs to offer some form of autoscaling that the team needs to build out. Approvals will be required to move an artifact into a production environment, especially when deploying during high-traffic times. The team will also need to figure out how to run the quality assurance tests and evaluate the results, enforce documentation and ticketing, validate that a production deployment was successful, and have some failure strategy.

The interviews that the DevOps team had done before will answer all of these questions. But now the team has to map out the sequence, check for integration requirements and API or CLI executions, and enforce security throughout the entirety of the pipeline.

Mapping is one of the most important and also one of the most boring steps that an engineer can go through. Not many engineers enjoy interviewing different teams, understanding requirements, and diagramming how the different relationships work together. However, if these requirements are not well understood and mapped out, then major gaps will be present in the final product.

To start the mapping process, it is important to gather the high-level requirements and map those first, then iterate upon that process map until relationships, permissions, and gaps are identified. Walking through these mapping iterations can be accomplished in the following ways:

1. **Solutions, platforms, and teams**: Understanding that continuous delivery is about getting every change out into an environment or to users in a safe, quick, and sustainable way is the desired outcome of the process. To achieve this outcome, a set of solutions, platforms, and teams must work together in a repeatable and reliable way. Understanding which teams, tools, and solutions currently exist and what gaps need to be filled is the first round of iteration.

   Platform: The easiest place to start the mapping process would be understanding what platforms are being used. The cloud providers or data centers that are being used and the orchestration of those platform resources are the two main areas to understand and map out.

   Tools: Every step in the continuous delivery pipeline is either controlled by a tool or solution or by a person. If complete automation of the pipeline is the desire, then any step that relies on a person is a gap that needs to be covered.

   Teams: Considering that every engineering organization has multiple teams involved in every release, understanding the role and requirement of each team is essential.

2. **Beginning the map**: With the platforms, tools, and teams understood, the next step is to build out the map.

   The platform section of the map can be represented as a pyramid stack, with the environment name at the bottom, the cloud provider or data center name next, and a virtual machine orchestrator, such as Vagrant, at the top:

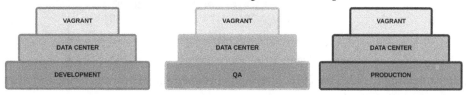

Figure 10.1 – Breakdown of each environment

The team section of the map can be represented as a person icon with an appropriate title underneath:

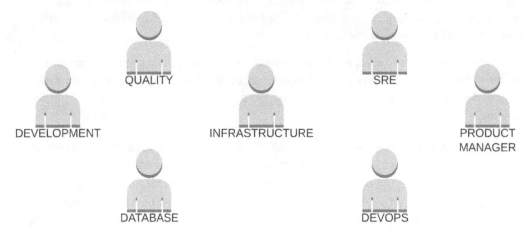

Figure 10.2 – Teams involved in the process

The tools section will be similar to the team section:

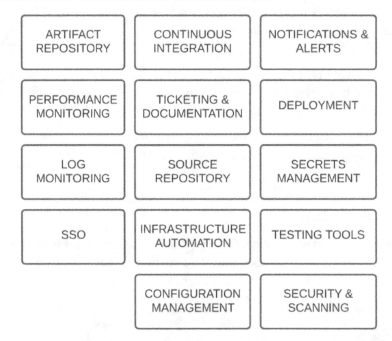

Figure 10.3 – Process steps

After mapping the high-level requirements between the platforms, tools, and teams, the next step is to highlight which areas are not automated currently:

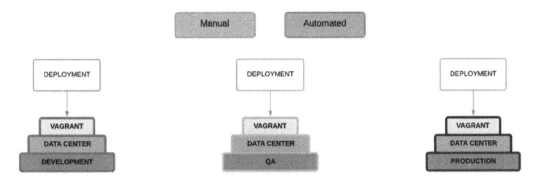

Figure 10.4 – Manual or automatic deployments

Then, to understand the current state of the delivery, every step of the delivery process will need to be tagged with either a manual, automatic, or partial execution type. This will help understand what can be or needs to be automated later:

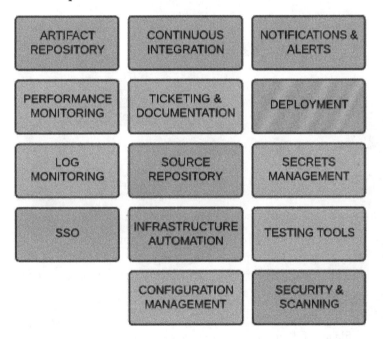

Figure 10.5 – Manual or automatic steps

3. **Pipeline process map**: With the tools, platforms, and teams defined, the last step in process mapping is putting together the pipeline process for the delivery and deployment:

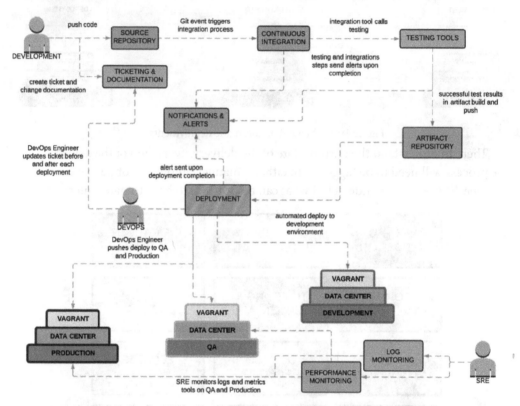

Figure 10.6 – Process map showing steps, teams, environments, and current execution types

The mapping process is manual and requires advanced knowledge of each phase, tool, and platform. If the team assigned to the process mapping does not have all of the information readily available, they will need to conduct interviews with other teams for better accuracy and map completeness.

After the mapping is complete, the team that is implementing verified GitOps will have to decide on the best file and folder structure for the declarative file breakdown. If the tool or solution being leveraged for verified GitOps does not have an automated file builder capability, such as Ansible or an in-house tool, then the team will need to decide whether they will have a small number of very long declarative files or a large number of very small declarative files. This decision is extremely important since it will be difficult, if not impossible, to restructure the files later. In many cases, the files and the accompanying solution will have to be rebuilt or completely reconfigured to switch from one file structure to another.

One manifest or many

The DevOps team had completed the process mapping and understood what was required for the delivery of each application type. They gathered information related to the tools, the platforms, the permissions and access requirements for the tools and platforms, and the teams that are concerned with each phase. If the engineering leadership decides to move forward with Ansible, then the DevOps team will need to convert the process map into the declarative file structure that Ansible expects. However, if the decision is to move forward with Harness, then the DevOps team will only have to build out the process map in Harness, which has its own automated file generation process on the backend.

Since the conversion of the process map to Ansible would be the most time-consuming process, the DevOps team decided to start building out the required Ansible playbooks. But, before they could start building those playbooks out, they had to decide on how the files would be structured. The main concern they had, related to structure, was to do with the length of the files versus the number of total files required.

It would be possible for the DevOps team to build out one very large file that contained a nested structure on what should be executed, allowing for the ability to skip certain steps. The benefit of the one-large-file approach is that every possible function or feature would exist in a single file and prevent any confusion associated with referencing multiple files. However, the main drawback for the DevOps team for this approach is that the file becomes unnecessarily long. The tool being used would have to reference the file for every execution and evaluate what pieces of the file are relevant to the specific execution. The alternative to the single-large-file structure is to leverage multiple smaller files that reference each other.

The DevOps team would have to make an easy-to-follow structure of folders to store the related files. Each file would require multiple file path references to achieve the desired functionality. The folder structure would make the configuration process simple, but any user that had not built the folder structure would have a difficult time knowing what files and folders had to be referenced to achieve the desired outcome. The many-files structure would require the DevOps to also build out an entire set of documentation for other users to reference when building out new execution pipelines.

After the team considered both of these options, they decided to use a hybrid of the two. This hybrid approach would result in multiple small files for the integrations and configurations and a few medium-sized files for the executions on each environment and the overall delivery process. The team would start with the files for the integration and configuration requirements, then move on to the larger files that would reference them.

Navigating through code files to understand what values and functions are being referenced and used is often called **spaghetti code**. In the spaghetti analogy, code files are compared to spaghetti noodles. If there is only one small noodle of spaghetti, then it is easier to untangle, similar to one code file being easy to understand. But as the code file grows beyond a manageable size, it can become difficult to understand, similar to a long spaghetti noodle becoming tangled with itself. The common solution is to take the large file and break it out into different categories, such as common functions and common variables.

Any tool can be easy to maintain and administrate when the execution and configuration behavior is simple. But, as more platforms, integrations, and commands are required, a tool can quickly become a heaping mess of spaghetti code. Some tools allow templatization through declarative files, while others are inherently imperative and have limited templatization capabilities. The declarative or imperative nature of the tool that is used will have a direct impact on the scalability, reliability, and usability of the tool. To avoid a mess of spaghetti code, it is important to understand how best to configure the folder and file structure as well as the reference capabilities of the tool. If the tool is imperative, then a small number of larger files would allow for less spaghetti code, whereas a declarative tool has more tolerance for a large number of smaller files.

To understand what spaghetti code is and how much of an issue it can be, here are some Ansible examples.

Simple LAMP stack deployment

This example is a simple LAMP (Linux, Apache HTTP, MySQL, and PHP) setup and deployment using Ansible:

Figure 10.7 – Example Ansible file and folder layout

The folder and file structure of this example starts with the group_vars folder, which is a folder that Ansible looks for to fetch group variables that can be applied to other files using templating characters such as { { and } }.

The roles folder allows the grouping of specific requirements and behaviors related to certain objects. In the preceding example, there is a common role, which can be applied to any object, a db role, which is specific for the database, and a web role, which is related to the web server that is being set up. Each role has a folder for handlers, which are actions taken when a change is made on the machine; tasks, which are what the role executes; and templates, which are what the role deploys.

The `hosts` and `site.yml` files are two files that live outside of the variables and roles folders. The hosts file tells Ansible where to execute the commands, and the `site.yml` file is the main Ansible playbook that is passed to the `ansible-playbook` command to run the playbook:

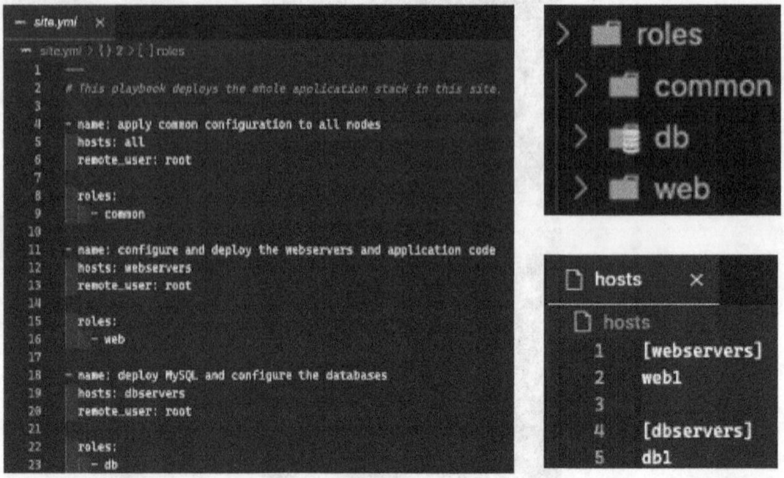

Figure 10.8 – Ansible file references

- The `site.yml` file starts by applying the common role configuration to every host and uses the root user to execute that command.

- The webserver role is applied to the webserver hosts, which come from the hosts file, using the root user as well.

- The `db` role is applied to the `dbserver` hosts, which also come from the hosts file, using the root user as well.

Templating allows variables to be passed into the desired commands using the `Jinja2` templating syntax:

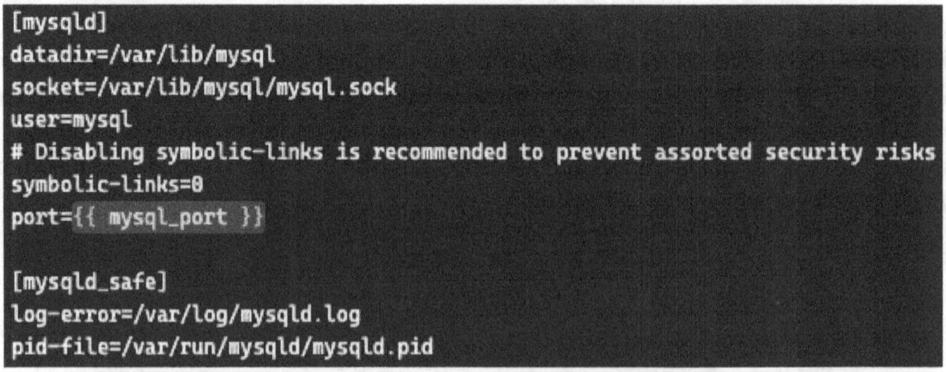

```
[mysqld]
datadir=/var/lib/mysql
socket=/var/lib/mysql/mysql.sock
user=mysql
# Disabling symbolic-links is recommended to prevent assorted security risks
symbolic-links=0
port={{ mysql_port }}

[mysqld_safe]
log-error=/var/log/mysqld.log
pid-file=/var/run/mysqld/mysqld.pid
```

Figure 10.9 – Template variable example

Install and Run – There are different methods of installing Ansible, all of which can be found at `docs.ansible.com`. Once Ansible is installed and configured, the command to run a playbook is `ansible-playbook -i hosts site.yml`:

```
(base) → simple git:(master) x ansible-playbook -i hosts site.yml

PLAY [apply common configuration to all nodes] *************************************************

TASK [Gathering Facts] *************************************************************************
```

Figure 10.10 – Ansible playbook run command

Advanced setup for Tomcat with memcached and failover

As the need for different capabilities and support requirements grows, so will the spaghetti code that is related to Ansible. In the case of needing a failover solution for clustered Tomcat, using NGINX as a load balancer, and Memcached as a session manager:

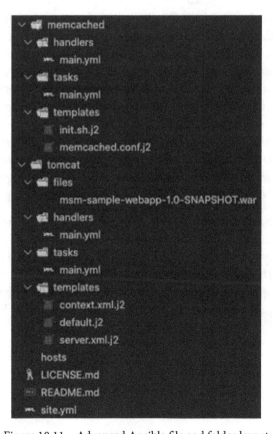

Figure 10.11 – Advanced Ansible file and folder layout

The folder structure is similar to the simple LAMP Ansible set structure, but there are significantly more required folders and files needed to achieve the desired outcome. For every stack or technology change, or as the platforms and infrastructure become increasingly more complicated and diverse, the spaghetti code can quickly grow out of control.

These two examples, which come from the public Ansible GitHub repository, are not as complex as what a multi-region or international organization would require. But, as can be seen from these examples, it is significantly easier to leverage something like Ansible to run the desired executions over building a tool in-house.

To scale a solution like Ansible to meet all of the requirements for a verified GitOps practice, a team will need to architect multiple layers of files to achieve a declarative file-based continuous delivery practice, especially when considering the required integrations.

A manifest for integrations

As the DevOps team started to build out the different declarative file structures for the application stacks, they decided that the easiest piece to start with was solution integrations. Some of the solutions were common across the application stacks, but other tools were unique to a specific platform or artifact type. An advantage to using an open source solution such as Ansible meant that some of the work might have been completed by others already. For example, the team needed an integration with a popular ticketing solution to handle some of the auditing requirements. Since the tool is widely used, it was highly likely that there are already plugins or examples of the tool being used in Ansible. After finding the community plugin and discovering how to add it to their Ansible configuration, they needed to figure out the best way to pass in the different requirements from the myriad of engineering teams and their differing processes into the tickets.

As the team continued to take inventory of the community plugins that were available in Ansible, they realized that most of the integrations they needed were already fulfilled by the community. A small set of solutions that needed to be integrated into the delivery process had no available plugin, which meant that the team would have to build out those integrations themselves. Once the plugins or scripts were built, the team would then be able to convert the unique operation requirements into declarative manifests that could be fed to Ansible at runtime.

The initial step is to build out the manifests required to accomplish the business and security needs for every delivery, such as auditing, security scanning, and approvals. Then the team would need to build out manifests for the specific requirements pertaining to each application stack. The last step is to configure example manifests that teams or developers would be able to use to add their own requirements to the delivery process. By the end of building out the integration manifests and integration scripts, the team would have an integrations folder filled with sub-folders and files that can be called for any and every delivery pipeline. After this is completed, they would then move on to declaratively defining the configuration management requirements for the different platforms and infrastructures.

Plugins are common ways for other users of a solution to contribute back to the community of users. Any tool that allows plugins to be shared and leveraged has the potential to become an extremely versatile and powerful tool. However, the balance between native functionality and supplemented functionality can make a solution either too rigid or too cumbersome.

A tool that has struggled with this balance over the past decade or more is Jenkins. **Jenkins** is an extremely powerful and versatile tool, but it can easily become overburdened with plugins, resulting in significant management and scaling difficulty. The over-extended Jenkins setup is well known and understood by the entire industry, and one reason for this issue is because of limited native functionality and an over-reliance on community plugins. However, the use of plugins does not make Jenkins a poor choice for engineering teams to leverage; rather, it is when an engineering team decides to rely almost entirely on Jenkins plugins that the issue arises. The same balance can be applied to a tool like Ansible.

Ansible has extension capabilities, similar to what is present in Jenkins. And the issues mentioned before, relating to Jenkins and plugins, are also possible with Ansible. Community contributors can build out a plugin for Ansible in Python, a common programming language, which allows certain commands to be run and some information to be passed from the user to the desired command. The combination of a plugin, Ansible YAML syntax, and the Jinja templating language allows for a declarative approach to a traditionally manual or imperative execution requirement. But the reliance on a plugin introduces risk to the platform, and that risk increases as the plugin becomes more essential to the delivery pipeline.

A good example of risk level compared to plugin essentiality is the use of a ticket creation plugin. A low-risk usage scenario would be allowing for a user to create a ticket via the plugin, but if the user does not have the correct information filled in, or if the plugin fails, the delivery pipeline reports the error but does not stop. Since the pipeline was not blocked by the failure, the risk to delivery is minimal. Inversely, if the business requires that a ticket is created for every delivery pipeline, then a failure of the plugin would halt all deliveries until the plugin is fixed. The risk associated with the plugin is significant, as it can block all deployments going out. Whenever a plugin is used, it is important to understand what it is accomplishing, what to do if the plugin fails, and how to test the plugin whenever it is updated.

An alternative to extensibility through a community plugin would be building out the integration internally. The integration can be built as a plugin and contributed back to the community, but the integration could also be a simple script execution that takes in any required variables. The advantage of building an in-house integration is the purpose-built functionality and the reliability and support provided internally. If the plugin fails because the integration endpoints have changed, the team internally will quickly be notified, can quickly research and fix any issues, and will also be able to implement any required workarounds.

The Ansible documentation has information related to using plugins or modules, building plugins and modules, and how to leverage YAML files for the plugins or modules. For a verified GitOps practice to be implemented, the integration for each tool should be defined in individual files, a values file should be leveraged for each use case, and the execution engine should use templating that takes in values from the integration file and the values file:

- **Integration file**: An example integration file for a ticketing solution needs to include enough information for the connection, authorization, and execution of the designated commands:

```
# Integration
- name: Ticketing Tool
  id: 'ticket.tool'
  uri: '{{ uri }}'
  port: '{{ port }}'
  username: '{{ user }}'
  password: '{{ pwd }}'
```

Figure 10.12 – Templated integration file

- **Values file**: An example values file for a ticketing solution needs to include any required fields that should be passed to the execution to successfully accomplish the ticket creation. Optional fields can also be leveraged for added capabilities and requirements:

```
# Create ticket
- name: Create Ticket
  ticket.tool:
    project: '{{ proj }}'
    operation: create
    summary: '{{ summary }}'
    description: '{{ description }}'
    issuetype: '{{ type }}'
  args:
    fields:
      user: '{{ email }}'
      status: '{{ status }}'
```

Figure 10.13 – Templated values file

The execution engine should be notified that it has to execute the ticket creation task. It will see the name of the task from the `Create Ticket` file and use the next line to designate which integration file should be referenced. The execution engine will get the connection information from the integration file, namely the integration endpoint and authorization information. Then, once the execution engine is able to make the connection, it will compile the variables from the rest of the create ticket step and pass them into the final execution command.

By leveraging the templating capabilities provided by Ansible, the integration files and values files are kept to a minimum number, allowing for increased scalability and repeatability by having the user provide any specifics at runtime. After the integration file requirements are built and added to the Ansible folder structure, the next step is to get any platform-specific configurations defined in code.

A manifest for configuration

The next set of manifests that the DevOps team would need to create relate to the configuration requirements for the different platforms that need to be supported. The current platform support requirements are traditional Linux servers, Kubernetes clusters, and serverless functions in a cloud provider. Each of these platforms has specific configuration requirements, mainly related to networking and scaling. The traditional server platform would require a configuration file to exist on the server, while also updating any application server requirements. The Kubernetes platform would require updates related to any support tools that live in the cluster, as well as any ConfigMap *or* Secrets *updates that might be required. In regard to the serverless function support, any configuration change will be associated with the deployment of the new function version.*

The first issue that the DevOps team encountered was that the changes to the server-based applications are randomly updated and dynamic in their size and scope. Although some basic configuration elements still exist across all applications and servers, the applications could vary wildly in their underlying configuration requirements. But one advantage that could be leveraged by the DevOps team was leveraging the Git repository to host two different configuration files, one being the core configuration that rarely changed and the other being the dynamic configuration file. When the developers wanted to execute a delivery, the execution engine would be able to trigger off of the Git repository event. This event trigger would remain the same for the Kubernetes and serverless-based applications as well.

With these configuration files being configured and pushed by the developers, the DevOps team would need to include a dry run and approval gate to allow the deployment to go out. Ansible has an easy way to give a user a dry-run output of a command and a hypothetical output as well. This will be helpful when needing to audit any changes and approve or reject them. Eventually, the team would like to get to an automated checking process that can evaluate the dry-run output and hypothetical changes for any issues before approving.

Once the team completes the process of building out the configuration manifests and getting Ansible to pull the configuration files from Git, they would then need to get Ansible to output the dry-run results and wait for approval before executing a deployment. The execution process would be the next piece to convert to code.

The previous section discussed the balance of using plugins or building out integrations when native functionality is either lacking or not present. Plugins prevent the user from having to build an integration but can result in major risk to a delivery pipeline. Building an integration requires a significant amount of investment in both time and effort, but it will reduce the overall risk. There is also a balance when considering the configurability of a system.

Most companies fit into one of two categories:

- Do what you want, but be successful.
- Do what we want, we are successful.

The first company type desires to give complete freedom of technology decisions to the developer teams. The teams can choose which platforms they use, which tools they use, how they operate and administer the technology, how they monitor the application, and even which programming language they build the application with. The trade-off for this level of freedom is that the team has to be successful based on pre-defined business metrics. There are some teams that thrive in a **wild west** style of DevOps, but when all administration, management, and support is delegated to the same team that should also be developing the application, the desire to have a central support group becomes ever more important.

The second company type prefers to have the developers focus on getting the application into the hands of users, without worrying about what tools or platform is required. The company will define the technology stack, hire only those with specified programming knowledge and experience, and employ a centralized DevOps team that focuses on the developers being unimpeded in their work. The drawback to a more opinionated approach is that developers have little freedom to innovate with their application functionality and feature set.

The balance between a highly configurable process and a highly opinionated process is another difficult line to walk. Most large companies operate in the highly opinionated style, but that is often the result of years of operation, strict auditing and compliance requirements, security requirements, and large software purchases. Many smaller companies or younger companies operate in a highly configurable process, simply because it allows them to operate at a higher velocity.

In the case of converting configuration requirements into declarative code, the balance between being configurable or opinionated will either increase or decrease the number of special cases that the delivery platform must support.

If the delivery platform needs to accommodate a more configurable style of operation and interactivity from the developers, then the developers need an easy way to input their configurations while also being prevented from breaking the process. But if the desired path is standardized and more opinionated, then the developers will only be able to supply limited configuration settings, such as memory or CPU allocations.

This is an example of a more configurable Ansible playbook that asks the user for a significant level of input to configure the execution:

```
- hosts: all
  input:

    - name: memory
      prompt: "Memory"
      private: no

    - name: cpu
      prompt: "Cpu"
      private: no

    - name: ticket_subject
      prompt: "Ticket summary"
      private: no

    - name: ticket_summary
      prompt: "Ticket summary"
      private: no

    - name: cluster
      prompt: "cluster"
      private: no

    - name: environment
      prompt: "Environment"
      private: no

    - name: artifact
      prompt: "Artifact"
      private: no

    - name: git_repo
      prompt: "Git repository"
      private: no

    - name: override_values
```

Figure 10.14 – Variables to be passed to templated files

The advantage is that the user has the ability to execute whatever they want, but the disadvantage is that the user must know all of the potential inputs and the correct combination to achieve the desired outcome. The alternative is a forced path that always has the same outcome because of the limited input.

One way to avoid the significant input at execution time would be to allow a user to inject a file that has any requirements or configurations already added to it. A default configuration file would also have to exist to fill in any variables that the developers would leave blank. Ansible has the ability to fetch these files:

```
- name: Store file into /tmp/variables/defaults
  ansible.builtin.fetch:
    src: https://github.com/company/application/defaults
    dest: /tmp/variables/defaults

- name: Store file into /tmp/variables/overrides
  ansible.builtin.fetch:
    src: https://github.com/company/application/overrides
    dest: /tmp/variables/overrides
```

Figure 10.15 – How to pull values files

Once these files are fetched, they can be passed to the execution for variable overriding and configuration management as needed. The process of combining the configuration files into the main execution files will be part of the execution behavior.

A manifest for execution

With the integration and configuration manifests built and all required plugins installed, the DevOps team was ready to move into one of the more difficult parts of continuous delivery: the pipeline execution. Each execution would leverage the integration and configuration manifests, as well as significant native Ansible capabilities. Some parts of the pipeline would need to run sequentially while other parts would be run in parallel. Approvals, health checks, and dry runs would all need to be defined in code as well.

As the team started building out the mockup of what the execution process should look like, they were informed that the engineering leadership team would be moving forward with purchasing Harness licenses. Although this meant that the team would have to start over with the integrations, the onboarding time should be significantly shorter. Harness has a file generation process that converts all of the setup in Harness to a set of YAML configuration files.

The DevOps team would have to start with getting the integrations configured in Harness, which were a few simple forms to fill out, and the integration configuration file was generated for them. They connected their Git repository to Harness, which allowed Harness to push the configuration files for the whole of the Harness platform to their Git repository. This allowed the team to either configure Harness through the user interface or through the YAML files in the Git repository.

After the integrations were configured, the team could then turn to the configuration files that the different platforms required. Some of the files were accessed by connecting Harness to the designated Git repository folder, while others had to be either uploaded into Harness or converted to Harness requirements. This was one of the main issues that the team had with a vendor solution like Harness, where the platform didn't allow for easy extensibility in every area they needed. The DevOps team had to figure out the best way to get the developers to put in their configuration requirements without requiring them to manually execute deployments through the Harness website. Although the Kubernetes and serverless platforms had significant support for overrides and configuration files, the traditional applications would require more work at the execution level to allow for easy configuration file overrides. What the team decided to do was create the variable overrides in Harness, which was converted to the configuration YAML files and stored in the Git repository, and then the developers would be able to add their variables to that file instead. It wasn't the ideal situation, but it would suffice for now.

Once the integrations and configurations were added, the team would then be able to move on to the pipeline process, which Harness calls workflows. Each workflow would require a pre-deployment stage that would add or change any platform-specific configurations. Then the workflow would include a deployment stage and a post-deployment stage, as needed. The DevOps team was able to build out these workflows pretty quickly, but they did not want to manage hundreds of workflows if they didn't have to. Similar to Ansible, where almost anything can be templatized, Harness can templatize the core information of the workflow, such as what is being deployed and where is it going. This templatization meant that the team could build out a small number of workflows that can be scaled across different artifacts and environments as needed.

As the workflows were created, tested, and changed, the team was able to see that the generated configuration YAML files were being affected as well in the Git repository. When the DevOps team wanted to make any major changes to the different Harness objects, they could make those changes in Git, and those changes would be reflected in the Harness website as well. Although it did take a few extra hours to get the integrations and configurations added to Harness, the building out of the execution steps was simple and scalable. The last thing that needed to be configured was complete delivery execution and a trigger that executed whenever the Git repository was changed.

Integrations and platform or application configurations are prerequisites for any successful deployment. Yet these precursors are useless without a process that leverages them. The execution process is a combination of integrations, configurations, pre-deployment steps, deployment steps, and post-deployment steps. And just as the balances discussed for the integrations and the configurations are important to consider and implement, there is a balance associated with the execution as well.

Any solution that is used for verified GitOps must have the ability to leverage both declarative language files and templating. The combination of these files and templating capabilities allows for the verified GitOps platform to scale as well, since even the execution pipelines are defined in code and should be able to take in variables. Templatizing the execution pipeline leads to the next balance consideration.

If an execution pipeline can be templatized, then there are conditionals that can be leveraged. A conditional is evaluating two or more different options and then choosing one of those options to move forward with. The ability to leverage conditionals makes a pipeline very powerful, since the behavior of the pipeline can vary wildly based on the values provided. Adding to the power of conditionals are looping capabilities, where a pipeline can be executed on a loop with different values for each execution, conditional skipping of steps in the pipeline, and even scaffolding of pipelines. A scaffolded pipeline is where the pipeline resembles the bones of a structured deployment, such as declaring that a certain pre-deployment and post-deployment step must happen to meet security requirements, but everything else in the pipeline is entirely configurable. The balance that must be considered and implemented is whether there are a small number of pipelines that are highly configurable or a large number of pipelines that are more rigid. A more pliable pipeline can introduce more overall risk, but it has low administration requirements, whereas a more rigid pipeline reduces risk but has higher administration requirements.

For example, if the administration team wanted to reduce the total number of pipelines, then they would have to build out a robust pipeline with every possible required step and ask the user to specify which steps are needed:

Figure 10.16 – Pipeline file with significant input options

Although an all-inclusive pipeline would result in reduced administrative work, there is significant risk in the deployment, since anyone would be able to run any of the tasks in the pipeline whenever they want.

Reducing the overall number of pipelines to a minimum is the best way to not over-burden the administration team. However, ensuring that the pipelines are not over-burdened with too many options will allow for better security and compliance while scaling. With this balance in mind, the last set of manifests to create are related to the full delivery, which links executions together to allow for sequential deployments of dependencies or of environments.

A manifest for delivery

As the DevOps team is building out the workflows for the delivery pipeline, they are trying to take into consideration the wide array of functionality that needs support. Although the basic platform deployments are easy to define and build, there are other teams that need to execute their process requirements as well. The cloud team wants to run their Terraform processes through deployment workflows, but they also want to provide some level of self-service Terraform executions for the developers as well. The quality team has different tests that need to be run as a part of the deployments. Security scans need to be run on the artifact before the deployment is executed, but only on the lowest non-production environment. And the auditing and compliance group requires that tickets are created and updated throughout the delivery, as well as having an approval gate before a deployment executes against production.

The DevOps team can add some of the peripheral requirements directly into the deployment workflow process. But other requirements, such as the cloud team's Terraform process, will have to be their own deployment workflow. Including some of the extra workflows into an entire delivery process is not necessarily difficult, but defining which workflow or pipeline to trigger based on requirements from the users that are provided at runtime might be difficult.

A true GitOps flow would be triggering the delivery when a change to a Git repository occurs. But the DevOps team would need to figure out which Git repository and file will need to be updated to indicate a deployment with Terraform versus a deployment without Terraform.

Outside of the triggering of the deployments, the ability to accomplish a continuous delivery pipeline is relatively simple. Harness can use and reuse workflows in a sequential and parallel pattern to achieve the desired results. The DevOps team would need to make sure that the appropriate approval gates were set and that the different teams had the correct level of permissions to avoid issues with the approvals as well. Once the pipelines were configured, as well as the triggers for the pipelines, the DevOps team decided to push a change to their application code to trigger the integration and delivery process from start to finish. Successful execution of the whole delivery pipeline would mean that the team would have achieved GitOps-based continuous delivery, at least until something changed.

A continuous delivery process, which entails being able to get any change to production in a safe, repeatable, and reliable manner, will have more complexity as the platforms, applications, and customers grow. The ability to create infrastructure in a repeatable and automated way will result in less idle time for developers and less downtime for end users. By incorporating automated testing and security scans into a delivery pipeline, there will be fewer bugs and less risk introduced into production. And by offering self-service capabilities associated with the creation or changing of underlying architecture, fewer engineers are hijacked from their productivity to accomplish basic and repeatable tasks. Developing the automation to achieve a desired outcome is not difficult. But making the automated process safe, secure, reliable, repeatable, and scalable is extremely difficult to accomplish.

A delivery pipeline should be seen as an orchestrator of orchestrators. The pipeline is an abstraction on top of the workflows, which executes the pre-deployment, deployment, and post-deployment steps. The workflow is an abstraction on top of the environments, artifacts, and integrations. Each of these layers of abstraction should be defined in declarative files and then made to reference each other. When a pipeline is finally built, it should be a file that only contains the references to workflows and any variables that need to be provided by the users:

```
pipeline:
- workflow: rolling
  workflowVariables:
     environment: development
     service: example-service
     infrastructure: dev
- workflow: rolling
  workflowVariables:
     environment: qa
     service: example-service
     infrastructure: qa
- type: APPROVAL
  properties:
     userGroups: administrators
     timeoutMillis: 86400000
- workflow: canary
  workflowVariables:
     service: example-service
     infrastructure: prod
```

Figure 10.17 – Pipeline file with reduced size

Each workflow is referenced by name and the variables are provided as well. The drawback to this structure is that it doesn't tell the user which steps are going to be executed, just the workflows that are run. The workflow will be longer than the pipeline code, since the workflow contains all of the deployment-related requirements. But the workflow will also have references to artifacts, environments, and integrations, which will have to be explored individually to fully understand what is being executed when the pipeline is triggered to run.

Properly implementing verified GitOps not only involves having all of the configuration code, infrastructure code, and application code stored in a Git repository, but also triggering delivery pipelines based on Git events. Most developers who submit code to a Git repository are familiar with the pull request process, which should be the event that triggers the integration solution to run a build and test process. The integration solution should finish a successful integration pipeline by building an artifact and uploading it into an artifact repository. Then, once the artifact is uploaded, the integration solution should submit a new file to the Git repository that includes the new version of the artifact. The file push event by the integration solution to the Git repository will be the event that triggers the relative delivery pipeline.

Every engineering organization will have a different approach to implementing verified GitOps, depending on how they choose to account for the different areas of balance mentioned in this chapter. But one of the opinionated parts of Harness is that the manifest size and number is predetermined for the company in many ways. This means that the way that multiple manifests are used to accomplish a desired outcome will be the same everywhere. Although some engineering organizations might want to implement a smaller-number-of-large-files approach to verified GitOps, it is important to see what verified GitOps in Harness can look like.

Verified GitOps with Harness

The DevOps team was almost finished with building out their Kubernetes-based applications in Harness. To do this, the team had to add the artifacts for their microservices as Harness Services, which combined their core Docker artifact, the Helm chart, and any overrides. After the Services were created, the team then added the different clusters they needed to deploy to, as well as the designated namespaces. Then they moved on to creating the different workflows that were required for deployments across the different environments. These workflows had to include any business, security, or compliance requirements for the appropriate environments. The DevOps team had to also include any steps required for pre-deployments and post-deployments from the developers, the quality engineers, and the cloud infrastructure team. After these were completed, the team created a full delivery pipeline that would deploy all of the workflows in sequential order, without any intervention by a user until the production deployment. The last thing to add was the trigger for the pipeline to execute.

A major benefit of using a tool like Harness is the built-in capabilities around notifications, failure strategies, and some of the verification benefits. The DevOps team was trying to figure out how to get an automated rollback working in Ansible, which would be triggered on either a timing limit or an error message. They also needed to consider the verification process after the production deployment, which the business mandated for the delivery process. The DevOps team had figured that they would work on the automation for that at a later time and instead have significant alerts and continually refreshed dashboards on their monitors to watch the production environment. Harness has an automated verification process out of the box, which made for an easy win for the team. The last piece, which was the easiest to solve, was the ability for different execution events to notify the appropriate teams and channels. Ansible, and almost any tool for that matter, has this ability. Harness ties well into their chat tool, can email their teams as needed, and can also send alerts to their incident management solution as well.

After achieving continuous delivery with their Kubernetes applications, the DevOps team would move on to the serverless and then the traditional applications next. With the triggering of the delivery pipelines coming from the Git repository event, the developers would rarely, if ever, access Harness or even press a deployment button. With all of the validation, reporting, standardization, and automation that Harness provides the company, they would have minimal administration time requirements moving forward.

Harness is structured in a multi-layer abstraction model, where basic building blocks are added and then combined to make larger objects. The structure of Harness looks similar to a pyramid.

Everything in Harness is based on a large number of native integrations, platforms supported, and templates. Those integrations, platforms, and templates can then be combined to make up the Services and Environments in Harness. The Workflows combine the two layers below it, as well as workflow-specific capabilities, into a set of pre-deployment, deployment, and post-deployment steps. Lastly, the Pipelines layer combines workflows, with optional approval steps, to create the final layer of abstraction.

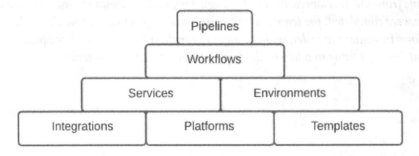

Figure 10.18 – Harness entity relationship

After signing up for a Harness account, one of the first steps will be to add an execution agent, known as a Delegate, to a location inside of the company network or **virtual private cloud** (**VPC**). The Delegate can be installed in a Kubernetes cluster, an ECS cluster, or on a Linux server:

```yaml
apiVersion: apps/v1
kind: StatefulSet
metadata:
  labels:
    harness.io/app: harness-delegate
    harness.io/account: sykvuk
    harness.io/name: test
  name: test-sykvuk
  namespace: harness-delegate
spec:
  replicas: 1
  selector:
    matchLabels:
      harness.io/app: harness-delegate
      harness.io/account: sykvuk
      harness.io/name: test
  serviceName: ""
  template:
    metadata:
      labels:
        harness.io/app: harness-delegate
        harness.io/account: sykvuk
        harness.io/name: test
    spec:
      containers:
      - image: harness/delegate:latest
        imagePullPolicy: Always
        name: harness-delegate-instance
        resources:
          limits:
```

Figure 10.19 – Delegate YAML

To install the delegate into the Minikube cluster, download the YAML from Harness and open the file in Visual Studio Code. Then you will want to change the memory resource requirements to 4 Gi:

```yaml
        name: harness-delegate-instance
        resources:
          limits:
            cpu: "1"
            memory: "4Gi"
        readinessProbe:
```

Figure 10.20 – Memory resource update

Once that change is done and the file is saved, while running `kubectl`, applying `-f` `harness-delegate.yaml` will install the Delegate into the cluster:

```
(base) →  ~ kubectl apply -f ~/Downloads/harness-delegate-kubernetes/harness-delegate.yaml
namespace/harness-delegate created
clusterrolebinding.rbac.authorization.k8s.io/harness-delegate-cluster-admin configured
secret/bf-test-proxy created
statefulset.apps/bf-test-sykvuk created
```

Figure 10.21 – Deploying the Delegate into Minikube

The Delegate will then execute all instructions that Harness gives it:

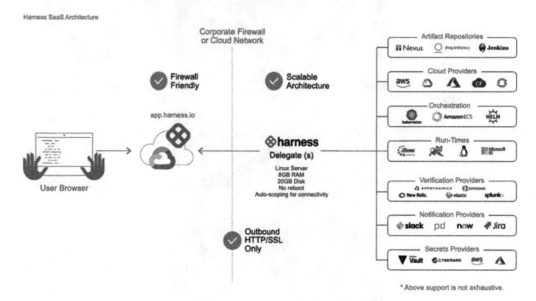

Figure 10.22 – Delegate connection

After the Delegate is installed, the next step is to connect to the required integrations and platforms using the appropriate connection credentials and permissions set. The first of these integrations is the Kubernetes cluster:

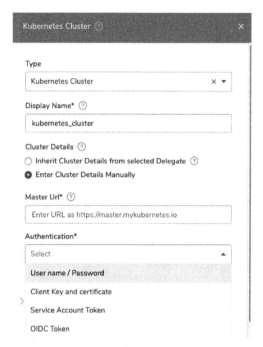

Figure 10.23 – Connect to Kubernetes cluster

After the Kubernetes cluster is connected, the next integration is the artifact server, such as the Docker Registry:

Figure 10.24 – Connect to the Docker Registry

The next integration is the Git repository, which can be connected either with SSH or HTTPS. You can use the same Helm chart repository as the last chapter did, but you will need to delete the secret and configmap from the chart first:

Figure 10.25 – Connect to Git repository

If there is a need for a ticketing or approval process through a collaboration system, such as Jira or ServiceNow, they will need to be integrated next:

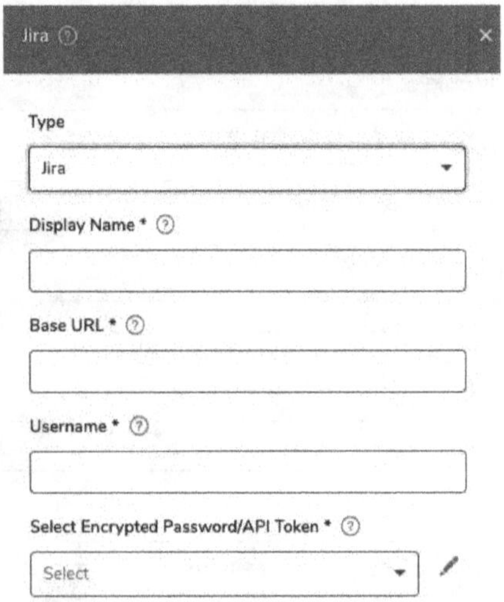

Figure 10.26 – Connect to Jira

Lastly, Harness has an advanced verification feature that can leverage the metrics from a monitoring solution, such as Prometheus:

Figure 10.27 – Connect to Prometheus

After setting up the integrations and platforms, the next step is to build out an application:

1. **Create an application**: An application in Harness combines all relative microservices, Environments, and so on.

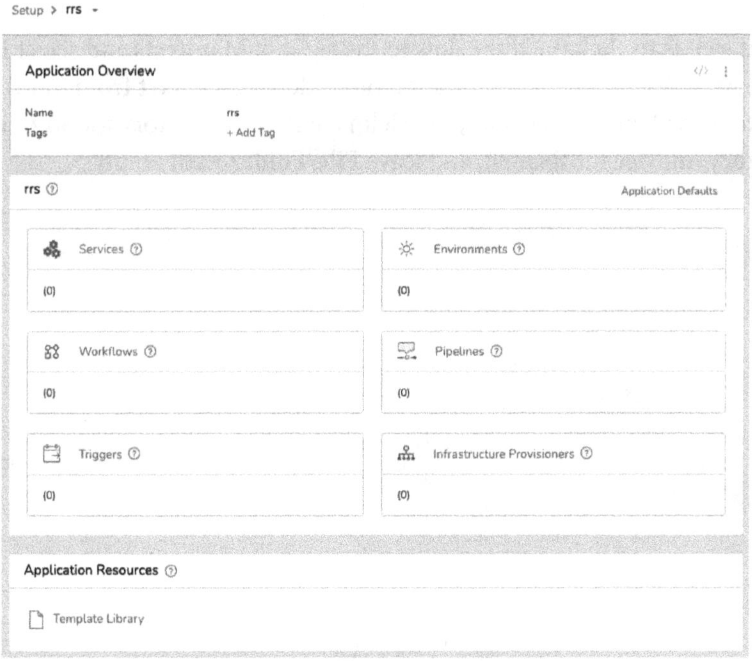

Figure 10.28 – Harness application

2. **Create a Service** – A Harness Service, in the example of Kubernetes, is the container artifact, the Kubernetes manifest, and any overrides:

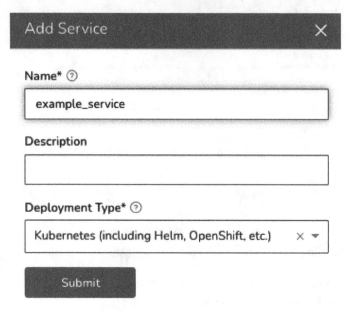

Figure 10.29 – Harness Service create form

The next step is to click the three dots to the right of **Manifests** and select **Link Remote Manifests**. In the configuration box, add **Helm Chart from Source Repository** for **Manifest Format**, the Helm chart Git repository for **Source Repository**, **master** for **Branch**, and leave **File/Folder Path** empty:

Figure 10.30 – Remote manifest setup

Once the information is added correctly, hitting **Submit** will save the configuration for the Harness Service to use:

Service Overview

Name	A-bf-helm-k8s
Deployment Type	Kubernetes (including Helm, OpenShift, etc.)
Artifact Type	Docker Image
Artifact Source	+ Add Artifact Source
Tags	+ Add Tag

Manifests

Manifest Format	Helm Chart from Source Repository
Source Repository	rrs (git@github.com:bfeuling01/rrs-book.git)
Commit	Latest from Branch
Branch	master

Figure 10.31 – Harness Service object

3. **Create an Environment:** An Environment is a group of infrastructures that can be associated with an access control grouping. For example, a development environment might include multiple namespaces in the same development cluster, as well as a set of development servers for a traditional application. In each Environment are infrastructure definitions, which define the infrastructure being deployed to:

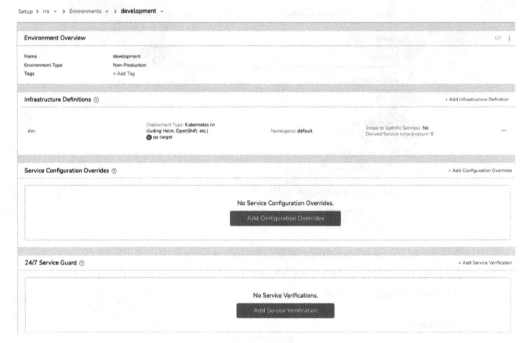

Figure 10.32 – Harness Environment object

4. **Create a Workflow** – A Workflow can include, but is not limited to, a Service, an Environment, and an Infrastructure Definition. A workflow will also include any pre-deployment and post-deployment steps that will be required when executing a deployment:

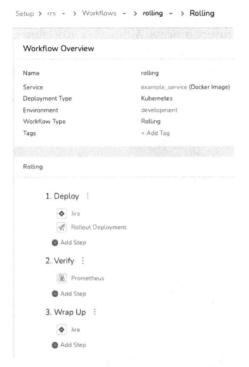

Figure 10.33 – Harness Workflow object

5. **Templatize the Workflow**: Every core object of a Workflow, namely the Service, Environment, and Infrastructure Definition, can be templatized. By templatizing the Workflow, it becomes a standardized deployment process that can scale across all Services and Environments:

Figure 10.34 – Templating a Workflow

6. **Create a Pipeline**: A Pipeline will combine Workflows and optional Approvals to accomplish a full delivery process. Although a Workflow can be deployed individually, if an engineering organization wants to execute a delivery pipeline across multiple environments, then a Pipeline is the best option:

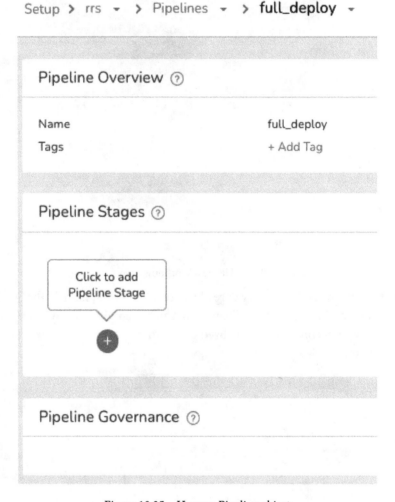

Figure 10.35 – Harness Pipeline object

After creating a Pipeline, the user can then add a new Stage, which can either be a Workflow or an Approval step:

Dev ✎ ⑦   ✕

◉ Execution Step   ○ Approval Step

Step Name*   ☐ Auto Generate Name

Dev

Execute Workflow*

rolling   ✕ ▾

Workflow Variables

Name	Entity Type	Value	
Environment*🔲	Environment	development	✕ ▾
Service*🔲	Service	example_service	✕ ▾
InfraDefinition_K... *🔲	Infrastructure Definition	dev	✕ ▾

☐ Execute in Parallel with Previous Step

Option to Skip Step ⑦

Do not skip   ▾

Submit

Figure 10.36 – Adding a Workflow to a Pipeline

A user can add as many Workflows and Approvals to a Pipeline as needed:

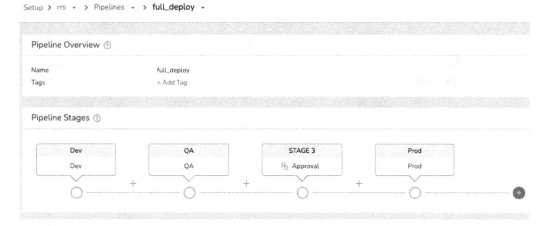

Setup > rrs ▾ > Pipelines ▾ > full_deploy ▾

Pipeline Overview ⑦

Name   full_deploy
Tags   + Add Tag

Pipeline Stages ⑦

Dev	QA	STAGE 3	Prod
Dev	QA	🔗 Approval	Prod

Figure 10.37 – Finished Pipeline

7. **Create a Trigger**: By adding a Trigger to the Pipeline, the Pipeline execution can be automated through an event. This allows the delivery pipeline to be executed in a continuous manner without any manual intervention requirements:

Figure 10.38 – Creating a Trigger

Once the Trigger is created, clicking the **Github Webhook** link will display the webhook for you to use in GitHub:

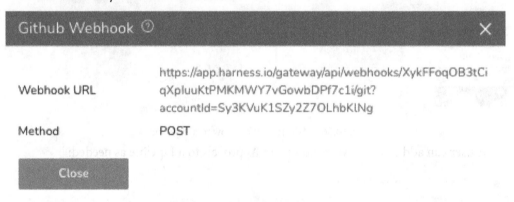

Figure 10.39 – Github Webhook

The webhook from Harness can be added into the Helm chart's Git repository by going to the **settings** page and clicking on **Webhooks**:

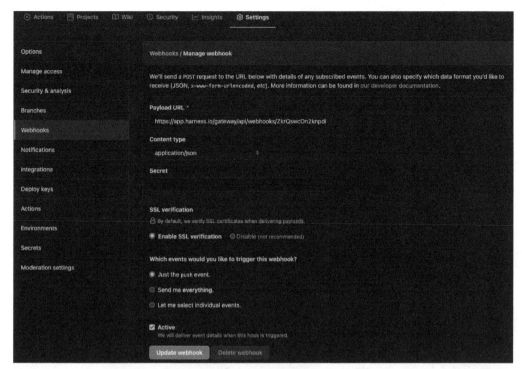

Figure 10.40 – Add the webhook from the Trigger to GitHub

After the webhook is added to the Git repository, a new file commit will trigger the deployment in Harness:

rolling (Direct Workflow Execution) - 02/15/2021 11:26 pm (less than a minute ago)

Application	rrs		Artifacts	None
Services	example_service		Instances Deployed	None
Environment	development		Infrastructure	dev
Workflow	rolling		Triggered By	full_deploy (Deployment Trigger

33 %

Pre-Deployment → Rolling → Deploy

Figure 10.41 – A new push event in the GitHub repository triggers the deployment

After anything in Harness is created, there is an automatically generated configuration YAML file. These configuration files can be accessed in Harness or by having Harness push the files to the designated Git repository:

1. **Configuration As Code in Harness**: The configuration code files in Harness are at the top right of the account setup page:

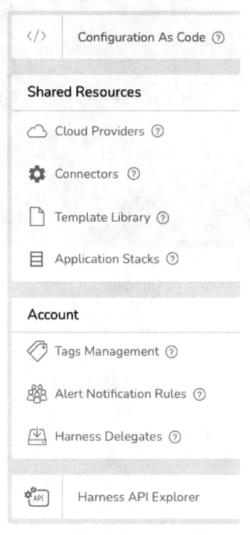

Figure 10.42 – Configuration As Code on the Harness setup page

In the Configuration As Code section of Harness, any file can be selected to show the code representation of that object, such as a Harness Service:

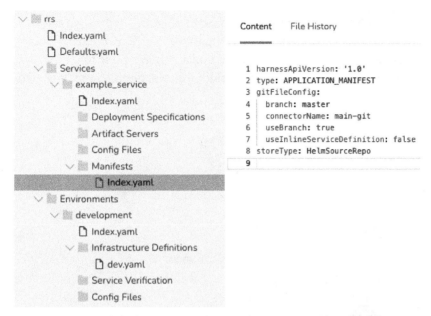

Figure 10.43 – Service configuration code file

Other types of files can also be viewed, such as cluster integration, the ticketing tool integration, or even the workflow YAML:

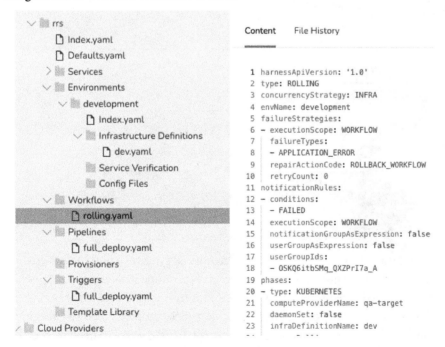

Figure 10.44 – Workflow configuration code file

2. **Setting up configuration code Git sync**: Setting up the Git sync for the configuration code is very simple. By going to the `Configuration As Code` section in Harness, you will see a file and folder hierarchy tree to the left. If you hover your mouse over an application folder, you will be able to see a Git icon:

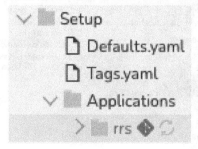

Figure 10.45 – Configuration code folder tree

When that is clicked, a popup will show the available Git repositories and the branch the configuration code should be sent to:

Figure 10.46 – Git Sync configuration

3. **Bi-directional Git sync**: To set up a bi-directional sync with GitHub, where a change to the configuration code in the Git repository will be pushed up to Harness, you will need to get the webhook link from the Git Repository connector in Harness. To do this, you will need to go to the **Setup** page, then to **Connectors**, and then click on **Source Repo Providers**, and, finally, you will need to click on the name of the repository connector that was set up previously:

Type	▼ Name	URL
Git	rrs	git@github.com:bfeuling01/rrs-book.git

Figure 10.47 – Git Repository connector

When you click on the name of the connector, a configuration screen will appear, allowing you to make changes as needed. On this configuration screen, there is a checkbox for **Generate Webhook URL**. Once checked and the configuration is submitted, a webhook link will appear. Add that webhook link to the desired Git repository's **Settings** page, similar to setting up the webhook for the Helm chart repository:

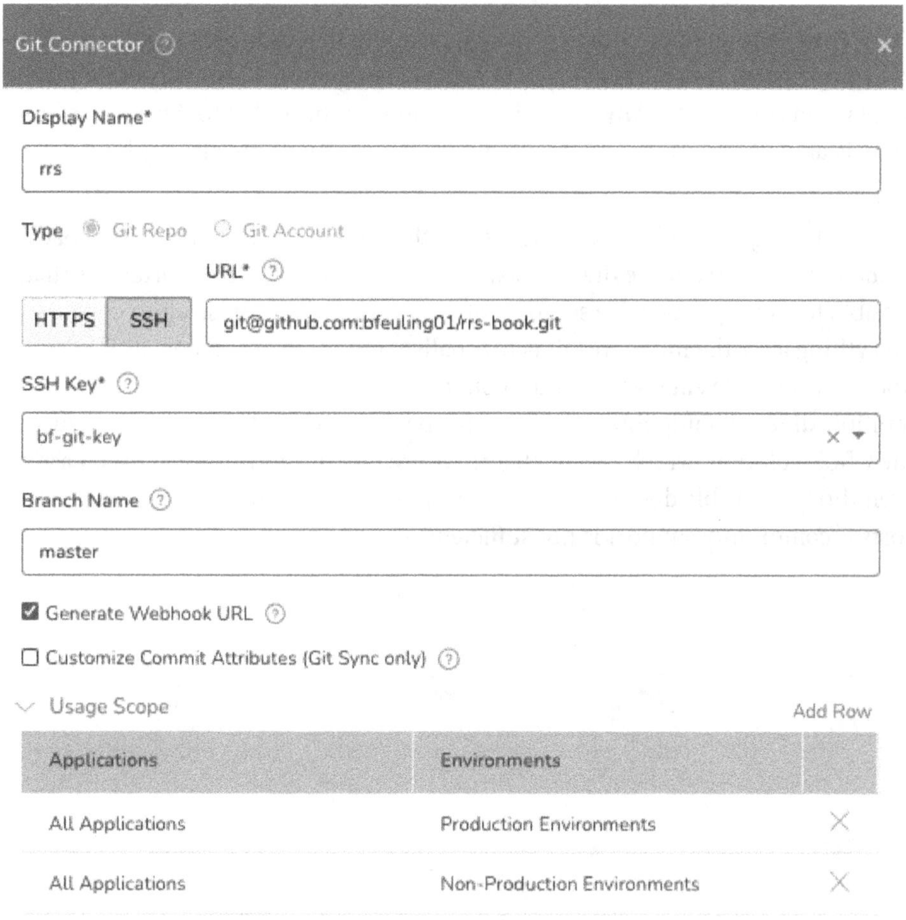

Figure 10.48 – Configure the webhook for the Git connector

The structure of files is similar in the Git repository, with the `Application` folder living in the **Setup/Applications** structure. Inside the `Application` folder are all objects belonging to the application as well. Other configuration code files that exist are the artifact repositories, the Git repositories, the verification providers, platforms, and so on. With every integration, platform, and every other object in Harness being automatically converted into declarative code, the ability to achieve a verified GitOps practice is easily attained.

When considering everything from this chapter, there are some important items to consider between Ansible and Harness, as well as best practices when adopting a verified GitOps practice.

Ansible:

1. **Pros**: Ansible is an open source solution, meaning that it doesn't require licenses to be purchased. It also has a major benefit of being almost infinitely configurable, since it can run scripts. Any desired integration can be added to Ansible, as long as there is an API or CLI. In fact, many of these integrations are already built out in the plugins.

2. **Cons**: Although Ansible is very extensible, there are some important drawbacks to the tool. The first of the drawbacks, and probably the most important, is that Ansible has a very expensive setup, configuration, and administration cost. Because everything after the initial install is manually configured, it can consume many hours for multiple engineers to build out a delivery pipeline that can deploy to multiple different endpoints. As with any open source solution, the users must host the whole solution, which means that there are hosting and scaling costs. Lastly, even though Ansible does not have a licensing fee, there are significant support costs if community support is not sufficient.

Harness:

1. **Pros**: Harness has a very light cost of ownership, both with regard to hosting and administration requirements. It also has significant customer support available and included through different chat options and a growing community. Lastly, the abundance of advanced capabilities and extensive integrations make for an easy-to-scale solution that provides confidence and reliability across all software delivery.

2. **Cons**: The drawbacks to using a tool like Harness are the common drawbacks associated with a vendor solution. The first, and probably most noticeable, is that there is a license fee to leverage the tool in a useful way. The other drawback is that the solution is closed source, which means that non-Harness engineers cannot alter the software code whenever they choose to meet their company's requirements. The only way to work around this issue is to either code the extensions themselves or to submit a feature request to the company.

There are no perfect tools, especially in the world of continuous delivery and GitOps. This is especially true when understanding that every tool will be implemented in a different way by every user and company. What is very important when considering a solution for continuous delivery and verified GitOps is to have an honest understanding of the benefits and drawbacks associated with the tools. Understanding both the pros and cons will aid in the proper analysis of which tool should be used and also what to look out for when implementing that tool.

Summary

This chapter discussed the different concepts related to building out verified GitOps using Ansible and Harness. Emphasis was placed on the different areas of balance that exist across manifest size and number, configurable and opinionated pipelines, and so on. Finally, the chapter went a bit deeper into how verified GitOps is shown in Harness.

After discussing originalist GitOps in depth in the last chapter, and verified GitOps in this chapter, it is important to show the pitfalls of GitOps. By understanding common issues that are encountered on the way to implementing GitOps, a team can properly navigate their GitOps implementation, which will be shown in the next chapter.

11
Pitfall Examples – Experiencing Issues with GitOps

A major section of this book was dedicated to covering the high-level understanding of different GitOps practices, how they are implemented, as well as benefits and drawbacks. This chapter is intended to show some of the drawbacks of GitOps in a more practical way.

To follow along with this chapter, you will need to follow the steps in *Chapter 9, Originalist Gitops in Practice – Continuous Deployment*, related to getting minikube, a Git repository, and Argo CD set up and running. The assumption throughout this chapter is that the user has Argo CD running in the minikube cluster and Argo CD connected to a Git repository. Although Argo CD is the main tool used for this chapter, that is only because Argo CD is the lightest and quickest to get set up and show pitfalls. This chapter is not intended to show issues with Argo CD, but rather pitfalls that can accompany GitOps in general.

In this chapter, we're going to cover the following main topics:

- Building and testing Kubernetes manifests
- Failure strategies
- Governance and approvals
- Proprietary manifest building

Building and testing Kubernetes manifests

With the DevOps team finishing out the migration of their Kubernetes applications in Harness, the business is starting to see significant improvements in the stability of their products. The next application for the DevOps team to add to Harness is the serverless application, and then the traditional applications last. Although the immediate requirements are to support the traditional applications alongside the cloud-native applications, there has been talk about the potential of decomposing the traditional applications into cloud-native structures.

For the traditional applications to move over to a cloud-native architecture, the teams will need to learn how to decompose, or slowly break down, the traditional applications into container-based functionality. The DevOps team will have to provide a set of documented best practices on how to get a container onto Kubernetes using Helm Charts. That will require that the developers not only have some understanding of a Helm Chart's structure but also understand how to add the different Kubernetes resources that their microservices require.

Some of the microservices will be customer-focused, requiring a more secure and easier-to-administrate network resource. Other microservices will be internal microservices, which need to be blocked from accessing outside endpoints and prevent external traffic from accessing the microservice directly. There will be microservices that require higher scalability, some will have a heavier resource consumption than others, and others will be shared services that many microservices in the cluster will interact with.

These different dynamic requirements will require significant build and test iterations of the individual Helm Charts to meet the desired outcome of the application teams. Developers will also need to keep their Helm Charts in a Git repository for tracking and versioning purposes, as well as for the DevOps team to help with any troubleshooting issues. But, as the DevOps team has worked through the process of building and testing Helm Charts before, they will need to show the developers how to access a development cluster and manually test the Helm Charts there, rather than through a GitOps tool. Whenever the DevOps team has tried building and testing Helm Charts through a GitOps tool, the workload would break and require some extensive work to fix it. Once the Helm Charts are working as expected, then the DevOps team would be able to add the Helm Charts into a continuous delivery pipeline.

Making a Helm Chart that fits a specific microservice's needs can be difficult. A common way to avoid this difficulty is to find a microservice with similar requirements that already has a Helm Chart built out for it. The engineer will copy the Helm Chart and tweak it to match their needs. An alternative to the copy method of Helm Chart building is to run the built-in Chart generation capability in Helm. The generated Helm Chart will have a decent Chart base for the user to build onto. Regardless of the method for building a Helm Chart, the engineer will need to test the Helm Chart with their service and then iterate on the Helm Chart until they have a stable Chart for continuous deployment.

Originalist GitOps, which is focused solely on deployment into Kubernetes, is not well suited for the building and testing phases of Kubernetes manifests. This is because originalist GitOps tools are intended to be the main mediator between the cluster and the user. Most originalist GitOps tools have an enforcement capability that forces the managed workloads to reflect the manifests in the Git repository. This means that if a user needs to tweak something with the workload and they go directly to the cluster to make the change, then the originalist GitOps tool will see the change on the cluster side and revert the workload to match what is in the Git repository. To understand this drawback, follow along with these steps (to ensure that the Helm Chart is well established, this Git repository shows a basic Helm Chart to work with: `https://github.com/bfeuling01/rrs-book`):

1. **Deploy the application** – With minikube, Argo CD, and a Helm Chart in a Git repository set up, you will need to create an application in Argo CD with the following configurations:

    -   **SYNC POLICY**: **Automatic** – When Argo CD sees a change in the Git repository, it will automatically synchronize what is in the cluster to match what is in the Git repository.

    -   **PRUNE RESOURCES**: Checked – If Argo CD has been deploying into the cluster and a new update of a managed workload manifest has removed a resource, then Argo CD will prune those resources from the cluster.

- **SELF HEAL**: Checked – If a change of a workload is enacted on the cluster side, rather than on the Git repository side, Argo CD will correct the workload in the cluster to match what is in the Git repository:

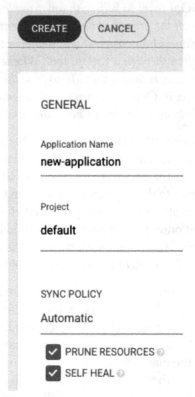

Figure 11.1 – Argo CD application with automatic sync, prune resources, and self-heal options

With those settings configured, the application will be deployed successfully:

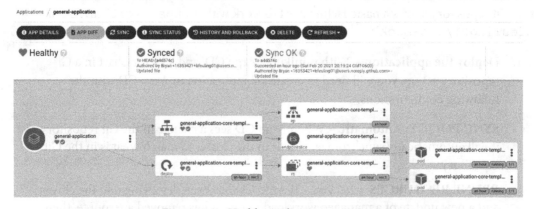

Figure 11.2 – Healthy application resources

2. **Change the workload in the cluster** – Access the cluster via your terminal
    and update the workload by editing the deployment using `kubectl edit`
    `deployment`:

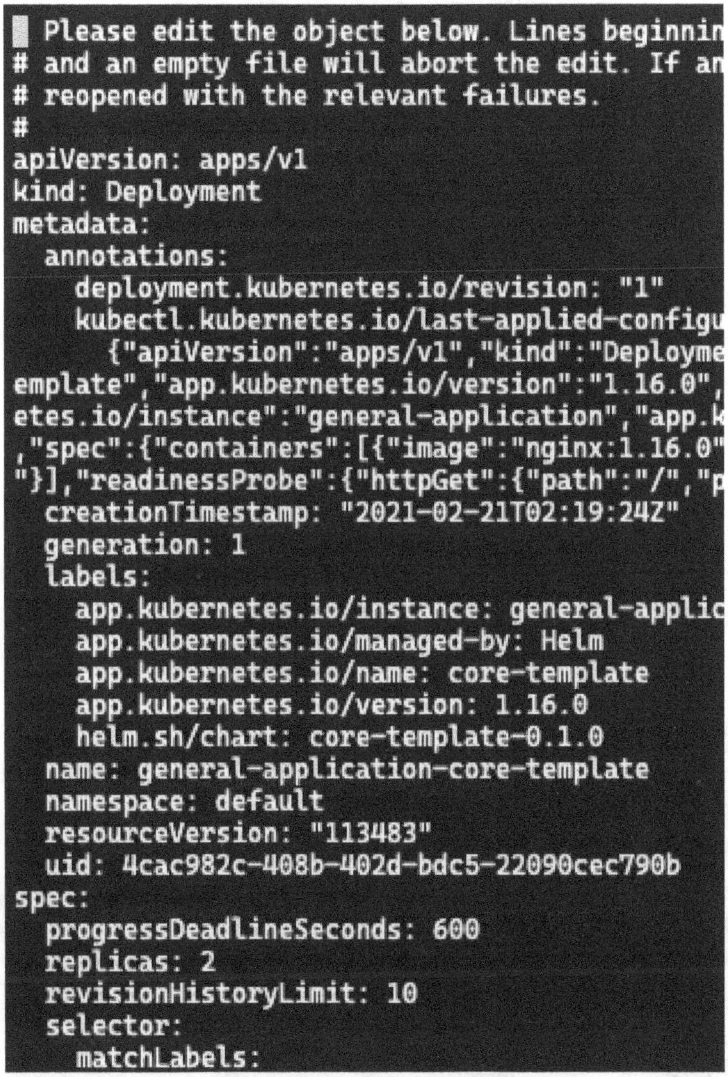

```
Please edit the object below. Lines beginnin
# and an empty file will abort the edit. If an
# reopened with the relevant failures.
#
apiVersion: apps/v1
kind: Deployment
metadata:
  annotations:
    deployment.kubernetes.io/revision: "1"
    kubectl.kubernetes.io/last-applied-configu
      {"apiVersion":"apps/v1","kind":"Deployme
emplate","app.kubernetes.io/version":"1.16.0",
etes.io/instance":"general-application","app.k
,"spec":{"containers":[{"image":"nginx:1.16.0"
"}],"readinessProbe":{"httpGet":{"path":"/","p
  creationTimestamp: "2021-02-21T02:19:24Z"
  generation: 1
  labels:
    app.kubernetes.io/instance: general-applic
    app.kubernetes.io/managed-by: Helm
    app.kubernetes.io/name: core-template
    app.kubernetes.io/version: 1.16.0
    helm.sh/chart: core-template-0.1.0
  name: general-application-core-template
  namespace: default
  resourceVersion: "113483"
  uid: 4cac982c-408b-402d-bdc5-22090cec790b
spec:
  progressDeadlineSeconds: 600
  replicas: 2
  revisionHistoryLimit: 10
  selector:
    matchLabels:
```

Figure 11.3 – Current deployment with two replicas

Alter the desired replica count, such as moving from two replicas to one:

```
# Please edit the object below. Lines beg
# and an empty file will abort the edit.
# reopened with the relevant failures.
#
apiVersion: apps/v1
kind: Deployment
metadata:
  annotations:
    deployment.kubernetes.io/revision: "1
    kubectl.kubernetes.io/last-applied-co
      {"apiVersion":"apps/v1","kind":"Dep
emplate","app.kubernetes.io/version":"1.1
etes.io/instance":"general-application","
,"spec":{"containers":[{"image":"nginx:1.
"}],"readinessProbe":{"httpGet":{"path":"
    creationTimestamp: "2021-02-21T02:19:24
    generation: 1
    labels:
      app.kubernetes.io/instance: general
      app.kubernetes.io/managed-by: Helm
      app.kubernetes.io/name: core-template
      app.kubernetes.io/version: 1.16.0
      helm.sh/chart: core-template-0.1.0
    name: general-application-core-template
    namespace: default
    resourceVersion: "113483"
    uid: 4cac982c-408b-402d-bdc5-22090cec79
spec:
    progressDeadlineSeconds: 600
    replicas: 1
    revisionHistoryLimit: 10
    selector:
      matchLabels:
```

Figure 11.4 – Deployment updated with one replica

After altering the replica count and saving the edit, the updated deployment will take effect:

```
NAME                                                    READY   STATUS              RESTARTS   AGE
pod/general-application-core-template-686bcf99d5-dftxw  0/1     ContainerCreating   0          1s
pod/general-application-core-template-686bcf99d5-n8jvh  1/1     Running             0          60m
pod/general-application-core-template-686bcf99d5-vb47f  0/1     Terminating         0          39s

NAME                                     TYPE        CLUSTER-IP      EXTERNAL-IP   PORT(S)    AGE
service/general-application-core-template ClusterIP  10.98.143.247   <none>        80/TCP     60m
service/kubernetes                       ClusterIP   10.96.0.1       <none>        443/TCP    25d

NAME                                                    READY   UP-TO-DATE   AVAILABLE   AGE
deployment.apps/general-application-core-template       1/2     2            1           60m

NAME                                                            DESIRED   CURRENT   READY   AGE
replicaset.apps/general-application-core-template-686bcf99d5    2         2         1       60m
```

Figure 11.5 – Kubernetes deployment updating the replica count

The moment that the updated deployment is saved, Kubernetes will look to alter the replica count. However, Argo CD will immediately revert the deployment to what the manifest says in the Git repository:

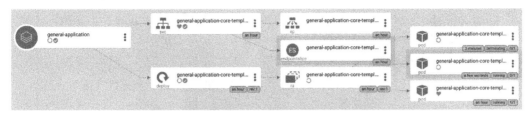

Figure 11.6 – Argo CD showing the outcome of the changed deployment and self-healing

Argo CD will show the updated deployment based on the in-cluster editing and also the outcome of the application being self-healed by Argo CD:

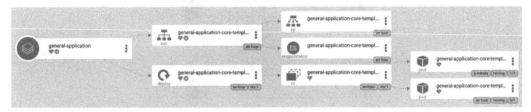

Figure 11.7 – Post-self-healing application health

3. **Change reverted** – Even though the change was made inside of the cluster, Argo CD auto-reverted the change to match the manifest in the Git repository.

The outcome of the change in the cluster not sticking around may cause some confusion for the end user that was looking to make small changes until they reached the desired state. But once they see that the change is automatically reverted, they will then choose to leverage the Git repository for their testing instead:

1. **Change the Git repository manifest** – To show a basic change in the Git repository taking effect in the cluster, change the replica count in the Helm Chart from two to one. Then, check Argo CD to see the desired outcome:

```
1   # Default values for core-temp
2   # This is a YAML-formatted
3   # Declare variables to be
4
5   replicaCount: 2
6
7   image:
8     repository: nginx
9     pullPolicy: Always
10    # Overrides the image tag
11    tag: ""
12
13    imagePullSecrets: []
14    nameOverride: ""
15    fullnameOverride: ""
```

Figure 11.8 – The Helm Chart values.yaml file

Changing the replica count will immediately cause Argo CD to sync the cluster automatically. To do this, go to the Git repository for the Helm Chart and make the appropriate changes in the `values.yaml` file:

```
1   # Default values for core
2   # This is a YAML-formatted
3   # Declare variables to be
4
5   replicaCount: 1
6
7   image:
8     repository: nginx
9     pullPolicy: Always
10    # Overrides the image
11    tag: ""
12
```

Figure 11.9 – The Helm Chart values.yaml file with the updated replica count

Argo CD will show the outcome of the updated Helm Chart after the effect takes place inside the cluster:

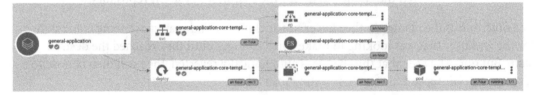

Figure 11.10 – Argo CD showing the reduced replica count

2. **Obvious breaking change** – Now that the replica count matches in the cluster, make a basic change that will break the deployment and watch the outcome in Argo CD:

```
1   # Default values for core
2   # This is a YAML-formatted
3   # Declare variables to be
4
5   replicaCount: 1
6
7   image:
8     repository: nginx
9     pullPolicy: Always
10    # Overrides the image
11    tag: ""
12
```

Figure 11.11 – Helm Chart with one replica

An obvious breaking change would be adding invalid information to the Helm Chart, such as a half replica count:

```
1    # Default values for core-
2    # This is a YAML-formatted
3    # Declare variables to be
4
5    replicaCount: 0.5
6
7    image:
8        repository: nginx
9        pullPolicy: Always
10       # Overrides the image tag
11       tag: ""
12
13   imagePullSecrets: []
14   nameOverride: ""
15   fullnameOverride: ""
```

Figure 11.12 – Helm Chart with an invalid replica count

Argo CD tries to deploy out the Helm Chart because of the updated files. However, because the deployment would be invalid, Argo CD shows an **OutOfSync** error instead of pushing the deployment down:

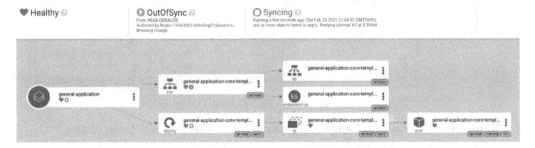

Figure 11.13 – Argo CD OutOfSync error

This change is an obvious breaking change, since there is no such thing as a half replica. Revert the change to get Argo CD to show healthy again:

```
1    # Default values for core-
2    # This is a YAML-formatted
3    # Declare variables to be
4
5    replicaCount: 2
6
7    image:
8      repository: nginx
9      pullPolicy: Always
10     # Overrides the image tag
11     tag: ""
12
13   imagePullSecrets: []
14   nameOverride: ""
15   fullnameOverride: ""
```

Figure 11.14 – Reverted Helm Chart with the correct replica count

When the breaking change is reverted in the Git repository, Argo CD will show a healthy application state:

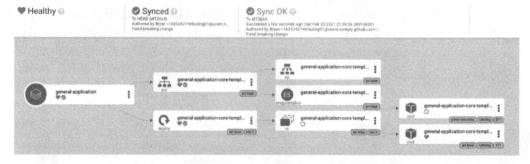

Figure 11.15 – Argo CD no longer shows OutOfSync

3. **Non-obvious breaking change** (volume mount) – The next task is to create a change that will break the deployment but make it non-obvious. This example will be using a volume mount that will not be found in the cluster. Updating `deployment.yaml` in the `templates` folder of the Helm Chart to include a bogus volume will show the following. As a result, Argo CD will show that the deployment is going through correctly, but what was added in the Helm Chart is wrong:

```
containers:
  - name: {{ .Chart.Name }}
    image: "{{ .Values.image.repository }}:{{
    imagePullPolicy: {{ .Values.image.pullPolicy }}
    ports:
      - name: http
        containerPort: 80
        protocol: TCP
    livenessProbe:
      httpGet:
        path: /
        port: http
    readinessProbe:
      httpGet:
        path: /
        port: http
    resources:
      {{- toYaml .Values.resources | nindent 12 }}
```

Figure 11.16 – Helm Chart with no volumes

Adding the volume and volume mount in the Helm Chart will trigger Argo CD to deploy:

```
    ports:
      - name: http
        containerPort: 80
        protocol: TCP
    livenessProbe:
      httpGet:
        path: /
        port: http
    readinessProbe:
      httpGet:
        path: /
        port: http
    resources:
      {{- toYaml .Values.resources
    volumeMounts:
      - name: config-vol
        mountPath: /etc/config
volumes:
  - name: config-vol
    configMap:
      name: log-config
      items:
        - key: log_level
          path: log_level
```

Figure 11.17 – Helm Chart with a volume mount

Argo CD sees the change in the Helm Chart and attempts to deploy it out:

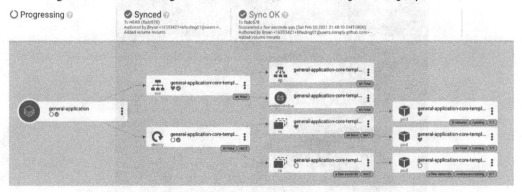

Figure 11.18 – Argo CD creating a secondary ReplicaSet to deploy out the replica

Argo CD does not remove the old deployment until the new deployment shows success. Because this never happens, there is a secondary replica set associated with the deployment in the cluster with bad resources:

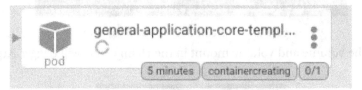

Figure 11.19 – Pod in the container-creating stage

As can be seen, the replica has been sitting in a waiting period for a long time so far, with no timeout period and no automated rollback. Argo CD can send notifications about the issue, but those would require individual notifications being set up for every application and Slack channel or email address. For the next step, correct the issue by reverting the change in the Git repository. Notice that the replica set, which is a resource that a Kubernetes deployment creates to manage a set of replicas, still exists for the failed deployment, but no replicas are associated with it:

Figure 11.20 – Abandoned resource

4. **Non-obvious breaking change** (no image pull secret) – The next, and probably more common, test is to deploy an image that is private and requires an image pull secret. However, the image pull secret will not be available for use and the container should have an error:

```
{{- with .Values.imagePullSecrets }}
imagePullSecrets:
  {{- toYaml . | nindent 8 }}
{{- end }}
containers:
  - name: {{ .Chart.Name }}
    image: "{{ .Values.image.repository }}:{{ .Values.image.tag | default .Chart.AppVersion }}"
    imagePullPolicy: {{ .Values.image.pullPolicy }}
    ports:
      - name: http
```

Figure 11.21 – Current Helm Chart

Update the `deployment.yaml` file in the Helm Chart to reference a specific container to which the cluster will not have access:

```
spec:
  {{- with .Values.imagePullSecrets }}
  imagePullSecrets:
    {{- toYaml . | nindent 8 }}
  {{- end }}
  containers:
    - name: {{ .Chart.Name }}
      image: 6515468712354.dkr.ecr.us-central-1.amazonaws.com/app:1.3
      imagePullPolicy: {{ .Values.image.pullPolicy }}
      ports:
```

Figure 11.22 – Container in a private repository

After specifying a private image, Argo CD will show an unhealthy application. This is because the replica in the cluster does not have the appropriate permissions to download the specific image:

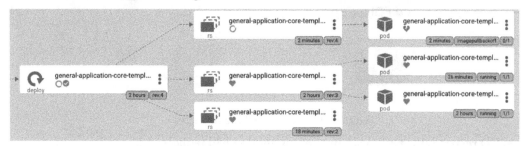

Figure 11.23 – Unhealthy application in Argo CD

Notice that there is one deployment, three replica sets, and three replicas. One of the replica sets is an abandoned resource, meaning that it does not have any replicas associated with it, nor is it referenced in our manifest anymore, but it still exists.

Another piece to notice is that although there are two pods that are running, there is a pod that is in a broken state. The two pods that are running are no longer represented in our Git repository, there is no automated rollback or cleanup happening, and now Argo CD is our source of truth, rather than the Git repository.

Deleting the replica set from the cluster can be accomplished through the Argo CD dashboard. In this case, selecting the option menu from the right of the Kubernetes resource reveals the **Sync** and **Delete** options:

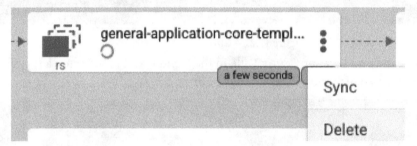

Figure 11.24 – Delete menu for the replica set

Once the **Delete** option is selected from the dropdown, the user is asked whether or not Argo CD should force the deletion of the resource:

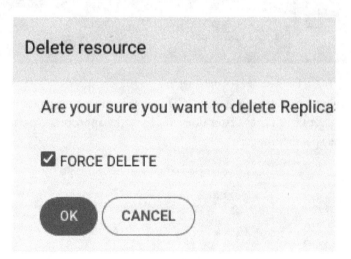

Figure 11.25 – Force deleting a resource

Argo CD recreates the replica set and creates another bad replica:

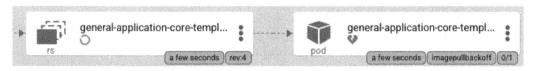

Figure 11.26 – Recreated replica set with a bad replica

Revert the changes to make the Git repository and cluster match, allowing Argo CD to be put back into a successful state:

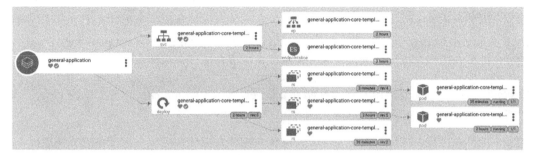

Figure 11.27 – Healthy application in Argo CD

Notice that there are now two abandoned replica sets in the cluster.

Building out Helm Charts is not always an easy feat to accomplish, often requiring multiple iterations to get the Charts correct. If there is a desire to move toward an originalist GitOps practice, beware of this type of pitfall, which can tie directly into automated failure strategies.

Failure strategies

A major requirement for the DevOps team is to have a way to trigger a rollback automatically when any issue occurs in the production environment. The company has a strict service-level agreement that specifies that their Software as a Service product has 99.99% uptime. Therefore, if their application is unavailable for more than 5 minutes a month, that would break the uptime requirement.

The team has researched the most common issues that would break a production deployment, how to test for them, and what they need to monitor to account for any other issues that might show up. The problem they have is how they would trigger a rollback in a GitOps model automatically.

They could try to leverage an automated revert command based on the outcome of the deployment status in Kubernetes. But that would require some significant scripting, leveraging a script, and figuring out a way to pass the specific repository that had triggered the deployment originally. However, even if they get that to work, they need to find a way to trigger a revert command based on the outcome of some alert from their verification process. The main reason they have to figure this issue out is that Kubernetes-based GitOps tools do not have an automated rollback capability. The only automated option that the tools have is an infinite retry loop, hoping that something would change that would make the tool show a healthy status.

Another failure strategy the DevOps team would like, although it is not a necessity for them, would be for a deployment to end the execution if there is a failure, allowing the team to analyze the failure to figure out the issue and prevent it from happening in the future. This failure strategy would only be applied to a lower-level environment but would be very beneficial for preventing issues in the production environment.

Regardless of what causes the issue in the production environment, the DevOps team has to figure out how to automate a failure strategy within an acceptable amount of time.

Any company that is looking to be successful and gain market share in the 21st century will require a highly available and highly desirable product. Although the highly desirable product is something that the business mainly concerns itself with, the availability of the product is placed almost entirely on the shoulders of the engineering team. Developers work to build a product that is functional and stable, DevOps engineers work to get the product from the developers to the end users, and production-focused teams, such as **Site Reliability Engineers** (**SREs**), work to keep production in a highly available state. And the best way to achieve all of these desired outcomes is to use automation and tooling to catch issues and execute appropriate remediation within seconds of a problem.

Developers, DevOps engineers, and SREs may all have different core responsibilities, but one process that links all of them together is **CI/CD**, or **continuous integration and continuous delivery**. Developers are impacted most by the CI process and SREs are impacted most by the CD process, with DevOps engineers being the responsible party for both integration and delivery. As a result of this, the capabilities of a CI or CD tool directly impact the ability of the engineering team to accomplish the desired reliability and scalability of a system.

One of the most important capabilities of any tool is the ability to return to a working state in an automated fashion. Infrastructure-related teams desire self-healing infrastructure, quality teams desire automated test analysis for flaky test alerting and for test cycle reduction, and delivery teams desire automated failure strategies, such as rolling back. Rolling back a deployment can have an immediate impact on the availability of the product or system. But if the rollback is not completed in the correct way or is not reverted to a known good version of the system, then the rollback can cause more harm than good.

Since GitOps is intended to have a Git-first mindset, where the Git repository is the source of truth, then a rollback is difficult because it must first start at the Git repository and not from the cluster or even the tool:

1. **Healthy deployment** – Make sure that the deployment is in a healthy state inside of Argo CD:

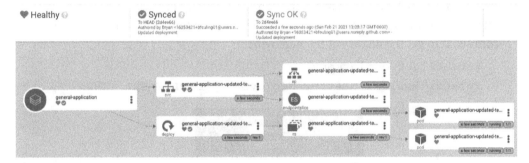

Figure 11.28 – Healthy application in Argo CD

2. **Introduce a breaking change** – Change the Docker image being deployed, as was done before:

```
spec:
  {{- with .Values.imagePullSecrets }}
  imagePullSecrets:
    {{- toYaml . | nindent 8 }}
  {{- end }}
  containers:
    - name: {{ .Chart.Name }}
      image: 6515468712354.dkr.ecr.us-central-1.amazonaws.com/app:1.3
      imagePullPolicy: {{ .Values.image.pullPolicy }}
      ports:
```

Figure 11.29 – Reusing the private artifact

Once the Docker image is changed in the `deployment.yaml` file for the Helm Chart, Argo CD will see the change and trigger the deployment. This will cause a bad deployment in the cluster:

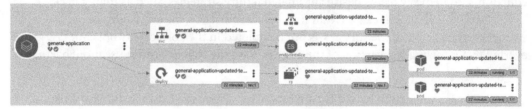

Figure 11.30 – Deployment is broken

A common way to fix a bad deployment in Kubernetes is to issue a rollback, which can be done through `kubectl` commands. However, because of how Argo CD is currently set up, a manual rollback will not have the desired effect on the cluster. To show this, you will first need to get the deployments that have been deployed out. Then, once the designated deployment name has been found, use the `rollout` command to get the history of the deployment:

```
kubectl get deployment
```
```
kubectl rollout history deployment <DEPLOYMENT NAME>
```

```
REVISION   CHANGE-CAUSE
1          <none>
2          <none>
```

Figure 11.31 – History of deployment revisions

To undo the rollout, leverage the desired `rollout undo` command on the deployment. Rerunning the `history` command will give evidence of the `rollout undo` command:

```
kubectl rollout undo deployment <DEPLOYMENT NAME>
```
```
kubectl rollout history deployment <DEPLOYMENT NAME>
```

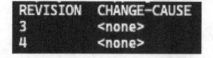

Figure 11.32 – Undoing rollout causes Argo CD to revert back to a broken state

Argo CD will show the broken deployment still being rolled out:

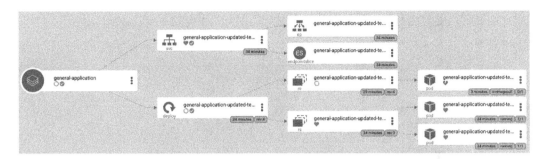

Figure 11.33 – Argo CD showing a broken rollout

Even though the rollback was executed within the cluster, Argo CD re-enforced the broken version in the Git repository. To overcome this issue, the "self-heal" option needs to be turned off.

3. **Turn off self-healing** – In Argo CD, go to the application's app details, edit the details, and turn off self-healing:

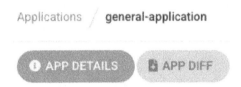

Figure 11.34 – Argo CD application details

Clicking **EDIT** will give the editable configurations for the application:

Figure 11.35 – Editing the application details

Different application details can be changed, such as disabling self-healing:

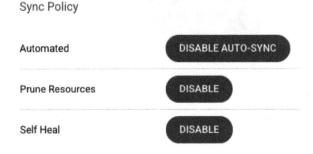

Figure 11.36 – Application automation features

Selecting the **DISABLE** button for the self-healing will lead to a confirmation screen:

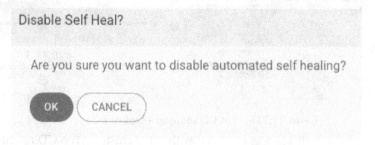

Figure 11.37 – Disable self-healing confirmation prompt

After selecting **OK**, the application details will show that the feature is disabled:

Figure 11.38 – Self-healing is disabled

With self-healing turned off in Argo CD, try running the rollback command in the cluster like before:

```
kubectl rollout history deployment <DEPLOYMENT NAME>
```

Figure 11.39 – Release rollout history

Running the `rollout undo` command on the Kubernetes deployment will trigger a rollback command:

```
kubectl rollout undo deployment <DEPLOYMENT NAME>
```

```
NAME                                                    READY   STATUS        RESTARTS   AGE
pod/general-application-updated-template-55f788fdb6-4xccf   0/1     Terminating   0          9m9s
pod/general-application-updated-template-96dcf9d57-4wfx5    1/1     Running       0          39m
pod/general-application-updated-template-96dcf9d57-z65k2    1/1     Running       0          39m
```

Figure 11.40 – Helm release rolling back

Notice that Argo CD is showing that the cluster and the Git repository are out of sync:

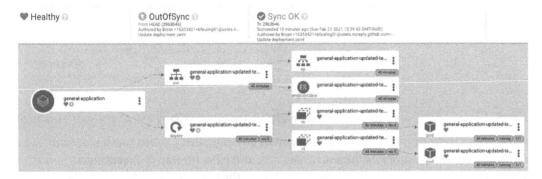

Figure 11.41 – Argo CD showing an OutOfSync error

To resolve the out-of-sync issue without causing a pod restart process, the user must update the Git repository to match what is in the cluster. If the user is not worried about the pod restarting process, then an update can be made to the Git repository directly, triggering Argo CD to update the cluster as needed:

```
28   28        {{- end }}
29   29        containers:
30   30          - name: {{ .Chart.Name }}
31   -            image: "{{ .Values.image.repository }}:{{ .Values.image.tag | default .Chart.AppVersion }}"
     31  +        image: 6515468712354.dkr.ecr.us-central-1.amazonaws.com/app:1.3
32   32            imagePullPolicy: {{ .Values.image.pullPolicy }}
33   33            ports:
34   34              - name: http
```

Figure 11.42 – Reverting the change to match the release in the cluster

With the update in the Git repository matching what is in the cluster, Argo CD will now show that the application is in sync:

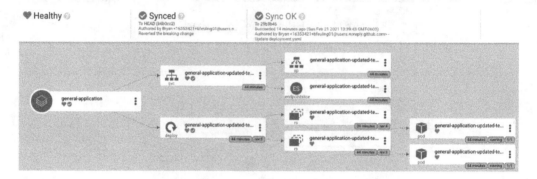

Figure 11.43 – Healthy application in Argo CD

This failure strategy limitation is consistent across all originalist GitOps tools, since the tools treat the Git repository as the source of truth and have no way of executing an automated rollback with an accompanying update to the Git repository to match the outcome of the rollback. As such, if there is a desire for automated rollbacks, then either the self-healing feature of the GitOps tool needs to be disabled and the rollback command must be executed upon an issue, or there needs to be a process of alerting when there is an issue, and a user reverts the change as needed.

If the rollback should be triggered automatically within the cluster, then a testing process needs to run, be evaluated, and then roll back when there is an issue. Another issue with releases is authorizing deployments into higher-level environments, such as through an approval or ticketing process.

Governance and approvals

As the DevOps team goes through the process mapping phase, one of the most important requirements that the company and the security team have is support for advanced governance and auditing throughout the delivery process. Because of the different data protection and privacy laws across the world, as well as financial auditing requirements, there are restrictions associated with user access, deployment approvals, interaction auditing, and execution auditing.

Every action and interaction of code from the moment that the code is committed to a Git repository until the code is in production must be tracked and audit-ready. Developers must create a ticket associated with the purpose for their work and documentation on what changed, and then submit the code and the ticket for peer approval before the code can be moved on to the integration process. Throughout the integration process, every test that is run and its output must be documented, as well as the outcome of the build. After the build is completed and the artifact is uploaded to the designated artifact repository, the delivery process is then triggered and either a new ticket must be created or the previous ticket is updated for a single activity record log throughout the integration and delivery pipeline. Information related to every environment, variable provided, test run, and approval given must be documented in the main ticket, which is usually done by different people throughout each stage. Once the artifact is ready to be deployed into the production environment, a pre-approved approver must evaluate the outcome of the different tests and deployments, then approve the artifact for deployment into production and update the ticket accordingly. Most of the execution and documentation steps are manual, resulting in significant bottlenecks throughout the entirety of the process.

Even with all of the documentation, checks, and balances being in place, the quarterly audits would still require the engineering teams to pause their work to assist with gathering audit information and resolving non-compliant issues. The ability to automate these compliance and auditing requirements, if possible, would both reduce the inconsistencies in the documentation and provide reliability at an increased deployment frequency. The team could potentially build the integration and process steps, adding them to containers and leveraging Helm hooks to update or create the tickets as needed through a GitOps tool. But once the DevOps team has found Ansible and Harness, they no longer need to focus too heavily on those issues, since both tools provide integration with their current toolset.

Almost every engineering team will need to submit their process and code to auditors at some point. Although the nature of the business and application will dictate the type of auditing that must be performed, an audit is required. The most basic audits that exist relate to data handling and the overall security of the system and architecture. But even in those basic auditing arenas, there are different levels of intensity for the audits. A **Personally Identifiable Information (PII)** audit is one of the most basic audits that exists and is related almost entirely to how information about a user is represented and presented to end users. **FedRAMP**, or the **Federal Risk and Authorization Management Program**, is the hardest audit process because of the large scope of controls, documentation requirements, third-party assessment organization review, and government entity authorization. Each of these audits has its own scope of requirements and audit cadence, whether that is monthly, quarterly, semi-annual, or annual. One of the easiest ways to avoid issues related to the auditing process is to gather all of the data points that an auditor would need, automate the gathering of that data, and then add that data to a single location with easy linking between the data and where it is found.

With regard to an originalist or purist GitOps process, since the process is deployment-only-focused, the compliance and approval requirements are out of scope for the native capabilities of the tools. However, that does not mean that there are no ways to achieve these requirements:

1. **Approvals and auto-sync in Argo CD** – Argo CD will need to have the auto-sync, auto-prune, and self-heal options disabled:

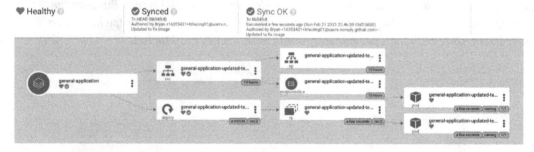

Figure 11.44 – Healthy application in Argo CD

Turning off auto-sync is done from the application details:

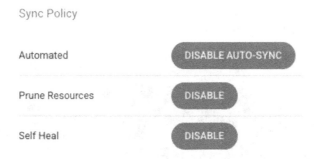

Figure 11.45 – Disabling auto-sync in the application details

2. **Approval through a pull request and auto-sync** – If auditing both who has access to the clusters and what actions are taken on the cluster is required, then leveraging the auto-sync and self-heal options of Argo CD will prevent users from making any permanent changes in the cluster. If the audit has an approval requirement for deployment, then originalist GitOps tools will have no native way to meet this requirement. Instead, the company could leverage the approval of a Git repository pull request.

Create a new branch, which is a Git-native way of creating a clone of the code to work with while avoiding any changes from affecting the main branch, in the Git repository by typing in the desired name of the new branch and selecting the **Create branch** option:

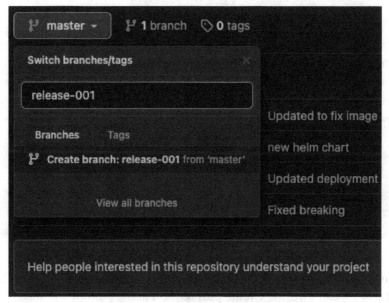

Figure 11.46 – Creating a new branch

Change the file in the new branch to show the name of the branch. This will provide a way to check that a deployment took place:

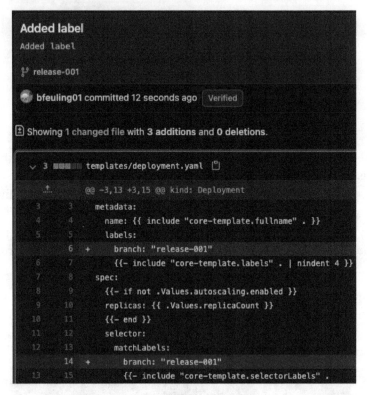

Figure 11.47 – Changing a file in the new branch

Once the change has been committed to the new branch, a pull request can be executed to merge the branches. A pull request is a way to take the changes in the code on one branch and add them to another branch through a merging process:

Figure 11.48 – Pull request is available

After clicking on **Compare & pull request**, an option to add any notes to the pull request will appear. Clicking **Create pull request** at the bottom will stage the new branch to be merged with the main branch:

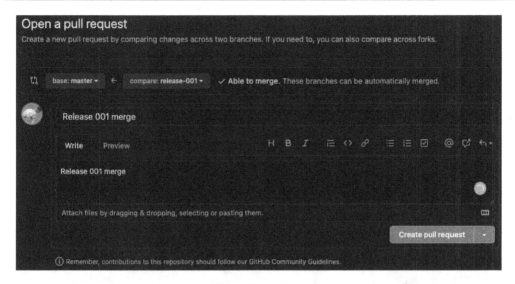

Figure 11.49 – Pull request

With the pull request being staged, **Merge pull request** will reconcile the two branches:

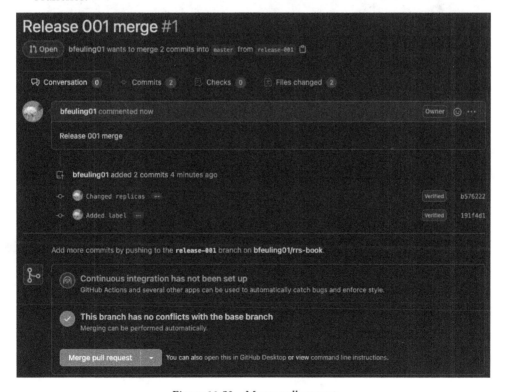

Figure 11.50 – Merge pull request

After clicking on the **Merge pull request** button, a merge confirmation page will show the recent action on the branches:

Figure 11.51 – Merge confirmation

With the merge complete, the main branch will now show the updated file:

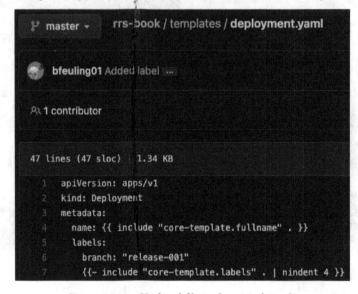

Figure 11.52 – Updated file in the main branch

The updated branch in the Git repository will be out of sync with the cluster, which Argo CD will show:

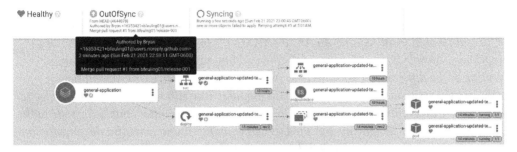

Figure 11.53 – Argo CD OutOfSync error

Since originalist GitOps is solely focused on deployment, any of the pre-deployment and post-deployment requirements, such as an approval step, must be executed externally to how Argo CD is triggered. A common way to solve these types of approval and testing issues is to have Argo CD be triggered in different ways based on a branching system. For example, if a user wants to test a new artifact, they might make a change to a development branch that triggers Argo CD to deploy into a specific cluster or namespace. Then, upon a pull request and merge to a different branch, such as QA, Argo CD will deploy the same Helm Chart and artifact to a different namespace or cluster than before. To fully automate this process, a user might leverage a complex system of ticketing and webhooks to generate pull requests and merges based on approvals.

But one area where originalist and purist GitOps have a major advantage is in the fact that the tools usually only require the core application manifests, such as a Helm Chart, to deploy successfully. Verified GitOps, which is delivery-focused, has a much more significant requirement with regard to manifest building.

Proprietary manifest building

One of the main pain points associated with building out any kind of manifest, regardless of language, is the requirement to learn the proprietary syntax. This is the biggest issue that the DevOps team has with a tool such as Ansible. Every action and execution that the DevOps team would want requires an Ansible YAML file to be created, stored, and linked appropriately. Ansible has made some of the file building easier by leveraging the naming convention of the different YAML objects as instructions. But when extra capabilities are brought into Ansible, such as plugins and extra modules, it would require extra knowledge to be added to the execution process. Also, adding a relatively uncommon templating style to the YAML files adds extra complexity. But the DevOps team has found extensive documentation and examples from the community to aid in this area, which gives them a step up when learning the nuances specific to Ansible YAML.

Harness has the same issues that Ansible has, with regard to the YAML file creation process. But adding to the confusion of proprietary manifest building in Harness is that there is little, if any, documentation about the Harness YAML syntax. There are also no Harness YAML validation tools to make sure that everything is valid in the YAML files. As a result of this, building out new workflows or services in Harness is nearly impossible to do from scratch through a YAML file. Harness supplements this issue with the fact that the user can build things in the application that are translated directly into a YAML file. This allows a user to create an example of what they want in the Harness application and leverage the resulting YAML syntax for building out templates later.

The DevOps team has not had any issues with how the tools perform and what the tools allow the teams to accomplish. But getting user adoption from the rest of the engineering organization would be difficult if they were to direct them toward a GitOps model and building files in code.

Originalist GitOps tools, such as Argo CD, have a much narrower scope of support and capability. This allows the originalist GitOps tools to reduce the user requirements to get fully configured and running. When the tool is quick to install and only requires that the user has the appropriate Kubernetes manifest, then the time to install is relatively short. This is not the case with verified GitOps tools, which require significantly more effort from the end users with regard to setup and configuration. Although an originalist GitOps tool can be heavily customized, the basic requirement is as simple as running the `install` command, setting up a connected repository, and letting the deployment happen.

If a user wanted to stand up a basic deployment of a Docker image to a Kubernetes cluster in Ansible, the file requirements would look like this:

1. **Deploy Helm Chart with Ansible** – The ability to accomplish a full delivery pipeline, where a Helm Chart is deployed across multiple environments sequentially, testing and security is included, and advanced deployment requirements such as Canary, is significantly longer to detail in an Ansible YAML. But the following code block is an example of a basic Helm Chart deployment into a default namespace:

```yaml
- name: Deploy latest version of helm chart inside default namespace (and create it)
  community.kubernetes.helm:
    name: test
    chart_ref: stable/busybox
    release_namespace: default
    create_namespace: true

# From repository
- name: Add stable chart repo
  community.kubernetes.helm_repository:
    name: stable
    repo_url: "https://kubernetes-charts.storage.googleapis.com"

- name: Deploy helm chart on 5.0.12 with values loaded from template
  community.kubernetes.helm:
    name: test
    chart_ref: stable/busybox
    chart_version: 5.0.12
    values: "{{ lookup('template', 'somefile.yaml') | from_yaml }}"

# From git
- name: Git clone stable repo on HEAD
  ansible.builtin.git:
    repo: "http://github.com/helm/charts.git"
    dest: /tmp/helm_repo

- name: Deploy Grafana chart from local path
  community.kubernetes.helm:
    name: test
    chart_ref: /tmp/helm_repo/stable/busybox
    release_namespace: default
```

Figure 11.54 – Ansible deploy latest helm chart version

If the version of the Helm Chart needing to be deployed is more specific, or requires dynamic variables, the YAML file becomes a bit longer:

```yaml
- name: Deploy latest version of helm chart inside default namespace (and create it)
  community.kubernetes.helm:
    name: test
    chart_ref: stable/busybox
    release_namespace: default
    create_namespace: true

# From repository
- name: Add stable chart repo
  community.kubernetes.helm_repository:
    name: stable
    repo_url: "https://kubernetes-charts.storage.googleapis.com"

- name: Deploy helm chart on 5.0.12 with values loaded from template
  community.kubernetes.helm:
    name: test
    chart_ref: stable/busybox
    chart_version: 5.0.12
    values: "{{ lookup('template', 'somefile.yaml') | from_yaml }}"

# From git
- name: Git clone stable repo on HEAD
  ansible.builtin.git:
    repo: "http://github.com/helm/charts.git"
    dest: /tmp/helm_repo

- name: Deploy Grafana chart from local path
  community.kubernetes.helm:
    name: test
    chart_ref: /tmp/helm_repo/stable/busybox
    release_namespace: default
```

Figure 11.55 – Ansible deploy specific helm chart version

2. **Deploy Helm Chart with Harness** – Harness has the same ability to deploy out Helm Charts, but the YAML looks significantly different. There are some core requirements for Harness, such as a Service, Artifact, and Environment, which can all be declared in YAML:

```
harnessApiVersion: '1.0'
type: DOCKER
imageName: library/busybox
serverName: docker


harnessApiVersion: '1.0'
type: APPLICATION_MANIFEST
gitFileConfig:
  branch: master
  connectorName: helm-charts
  useBranch: true
  useInlineServiceDefinition: false
storeType: Remote


harnessApiVersion: '1.0'
type: INFRA_DEFINITION
cloudProviderType: KUBERNETES_CLUSTER
deploymentType: KUBERNETES
infrastructure:
- type: DIRECT_KUBERNETES
  cloudProviderName: default-cluster
  namespace: default
  releaseName: release-${infra.kubernetes.infraId}
```

Figure 11.56 – Harness Service YAML

Once the different core requirements are configured in Harness, then the deployment definition can be created, which links all of the previous core requirements into a longer YAML file:

Figure 11.57 – Harness workflow YAML

Although the scope for both Ansible and Harness is significantly larger when compared to Argo CD and other originalist GitOps tools, the accompanying configuration requirements can also be extensive. Ansible and Harness have ways of mitigating these configuration requirements, whether that is auto-generation of the files, such as in Harness, or extensive templating capabilities.

Summary

By understanding the pitfalls that can occur, regardless of which GitOps practice is implemented, the team can better consider what will need to be done to avoid issues. Another major benefit of understanding these pitfalls is the ability to accurately understand how a tool or practice achieves the desired outcome, and what the required effort will be.

The next chapter will be a summary chapter that talks through what the book has covered, lessons learned throughout the book, and what the next steps look like for those wanting to implement GitOps in their company and organization.

12
What's Next?

This book has covered many different topics in the DevOps space. Although there has been a particular focus on GitOps practices, the overall concepts are geared toward the three main automation requirements of reliability, repeatability, and scalability. The best way to achieve the three automation requirements is through the practice of converting all operations into declarative code, leveraging a templating language, and enforcing the usage of a code repository with versioning capabilities. The goal of this last chapter is to provide an overview of some main concepts from the book and what the best next steps are.

In this chapter, we're going to cover the following main topics:

- Delivery versus deployment
- GitOps: what and why
- Continuous deployment GitOps: originalist and purist
- Continuous delivery GitOps: verified
- Best practices first, then GitOps

Delivery versus deployment

The DevOps team was continuing to onboard their different technology stacks and applications into Harness. They had been making great strides in the process of templating different requirements, enabling and training the other engineers on using Harness, and moving all of the container and serverless applications through a delivery pipeline.

Although the process was moving smoothly, there were some engineers within the company that expressed hesitation with a tool like Harness. Most of the time their concern was related to their lack of familiarity with the tool, or their previous experience of success with a different tool. But, in all cases of concern, a fundamental recurring issue was present: a lack of clarity about what deployment is versus what delivery is. The DevOps team had to go through the same learning process internally, which meant that this lack of clarity came as no surprise to them. But what they needed to decide on was whether they should add this difference to their enablement process, or if they should keep their focus only on training the teams on Harness. At the very least, they need to include a high-level understanding of each, mainly because without a tool like Ansible or Harness, the teams would have to manually fulfill the delivery pipeline requirements that a deployment does not cover.

Understanding the difference between delivery and deployment is essential to a successful software development practice. Whether an application is deployed hourly or monthly, or if the application is customer-facing or not, the application must pass through a delivery and deployment pipeline. For many companies, the delivery pipeline is mostly manual, requiring a significant amount of user interaction. These manual steps are most commonly found around testing, approvals, compliance requirements and ticketing, and even verification. Manual tasks will always result in bottlenecks, slow-downs, infrequent deployments. Conversely, the most common tasks that are automated are related to the deployment of new code. This is because deploying the code is usually a single command that needs an automated trigger.

At its core, a deployment is the process of getting any artifact, which is the application code, into an environment. When the application is monolithic in design, meaning that every feature and function is packaged together, versioned together, and is tightly coupled to each other, so the process of deployment can be significantly more intense. Monoliths can require significant configuration changes, often have longer shutdown and startup times, and most issues with a new version of the code will require hot fixing rather than rolling back. Often in contrast with a monolith is a microservice, which is significantly smaller in size and functionality and requires less deployment-related effort. A microservice is a function or subset of functions that are packaged together and typically interact with other microservices to achieve a full featured application. A microservice architecture is loosely coupled by design, and the typical deployment effort is often minimal, requiring a single command. But, with both monoliths and microservices, a deployment is only the way of getting the artifact into an environment. This means that a deployment pipeline must be available for every artifact and every environment. And as the different environments have different requirements in regard to technology stacks, configuration changes, and user interaction, many of the pre-deployment and post-deployment requirements can cause significant bottlenecks and timing issues.

If a deployment is focused on the artifact getting into an environment, then a delivery is all of the pre-deployment and post-deployment steps associated with getting an artifact from its initial building to the end user or a production environment. The delivery pipeline will often include ticketing or documentation building, notifications associated with the execution process, testing of the new code and artifact, security scanning, auditing and compliance requirements, approvals, verification and validation of the new code, failure strategies, scaling strategies, and progressing an artifact through multiple environments. A delivery pipeline can vary wildly based on the requirements of the company, the type of application, and the scale of desired endpoints. The importance of the application to the business will directly impact the number of requirements that must be included in a delivery pipeline, even to the extent of requiring multiple approvals from different teams and business units to deploy to production. Another major factor of the delivery pipeline behavior is the compliance, security, and auditing requirements that a company must adhere to. Any company that must submit to PCI, or payment card industry, compliance will have a significantly different pipeline than a company that must abide by **HIPAA**, or **Health Insurance Portability and Accountability Act**, standards.

The differences between delivery and deployment are essential to understand; delivery and deployment are often mistaken for one another. In many cases, companies think that they have delivery figured out and automated, but they only have their deployments covered. A common example of this is when a continuous integration tool executes the deployment command at the end of the build process. But getting from the artifact being built to the deployment in production often requires multiple teams or people and long wait times. When a company is only focused on the deployment, then their employees must bear the manual requirements for the rest of the delivery requirements. And manually fulfilled delivery requirements will always lead to insufficient effort, insufficient information, or missed deadlines, or a combination of the three.

GitOps – what and why

One of the major issues when adopting any new process is the ability to implement that process effectively and efficiently. The importance and impact of the process will directly correlate to the amount of time and effort required. But for the DevOps team, they were needing to implement both a GitOps practice and a new tool. And to add to the difficulty of a parallel implementation process, the team had to figure out the best scaling process for their applications.

In the case of Kubernetes and Hem charts, one of the major issues is manifest sprawl, which the DevOps team had experienced before. Their current hurdle was figuring out how best to manage and maintain that manifest sprawl when many of the applications are moving to containers and Kubernetes. The typical solution is to have a Git repository of many different Helm charts, one for each microservice, and have the teams maintain them. But that explosion of Helm charts would bring with it a massive troubleshooting and configuration issue. There is no good way for the DevOps team to maintain, understand, and enforce standards on every single Helm chart. But one of the advantages to their new continuous delivery GitOps method is that the team can specify any Helm chart and any override file combination that they want to, especially if the Helm charts and override files exist in a Git repository. This meant that the DevOps team could consolidate all of the different Helm charts into a small subset of Helm charts, or maybe even a single core chart, because the override files would dictate the Helm chart behavior. With the Helm chart consolidation capability, the only thing that the engineers would need to maintain is a small override file, which would be similar to something that the teams were already used to.

The other side of the scaling problem is related to scaling each microservice and application for each team onto the new GitOps tool. Although Harness has configuration code capabilities, there is no capability to leverage a single template across multiple applications. This meant that the DevOps team would need to build out a way for their applications to be automatically added with little to no effort, especially as the applications and teams grow. And since Harness has the ability to send the configuration code files to a Git repository, as well as having changes to the files in the Git repository be sent back to Harness, the DevOps team needed to only find a way to create a new set of files, commit them to the Git repository, and send them up to Harness. And because tools such as Ansible and Harness have a declarative nature to them, many of the steps can be templatized, allowing for easy replication and scaling across different applications. The scalability and repeatability that comes from the templating and configuration code file capabilities were the main reasons why GitOps was of such importance to the DevOps team.

GitOps as a practice has a different definition and application, depending on who is asked. Some dictate that GitOps is treating the Git repository as the source of truth. Others will say that GitOps is about converting every operation step into code, which is stored in a source code repository. And there are some that will give a middle-ground hybrid of the two. But at the core of every GitOps definition is the same basic principle, which is code. To be more specific, every action or desired outcome of an operation should be defined in declarative code files, with the key term being **declarative**.

Declarative code files mean that the files are not giving the actual instructions related to what should be happening, but rather just what the outcome should be. The opposite of a declarative approach is an imperative approach, which contains both the desired outcome and the process of reaching it. These two approaches will always work in tandem, since a declarative process is sending the desired outcome to the imperative process. But the extensibility of an imperative process allows for broader declarative capabilities to be leveraged. An example of these two would be the process of ordering a pizza. Most customers will not want to call a restaurant and tell them every individual step related to building, customizing, and baking a pizza. Rather, the customer will want a simpler process, which is typically just telling the restaurant what type of pizza is desired. Although there is an imperative process currently in place at the pizza restaurant to make and bake the pizza, the restaurant will want to provide their customers with an easy way to order the pizza. To take the analogy a bit further, imagine if the pizza restaurant only allowed for cheese pizzas to be made. The customers would not have many choices, but then they would also have to put in less effort when ordering. But if a pizza restaurant offered 50 different topping options and 4 different sizes, then the customers would have more to choose from and therefore more information to give. The customer ordering is the declarative part of the process, since they are declaring what they want. And the restaurant has the ingredients, recipes, and baking directions to make the desired pizza, which is the imperative part of the process.

As with the analogy, a GitOps tool can be very limited in what it supports, which leads to fewer options and fewer declarative requirements from the user. Additionally, a GitOps tool can be broad in what it supports, which leads to more options and more declarative requirements for the user. There is no one-size-fits-all configuration that can be applied to every engineering team, but it is important to understand how opinionated or configurable a tool can be. If the company requires that multiple container orchestrators are supported, then a Kubernetes-only GitOps tool would not be beneficial. But if everything is already automated and the team is looking for a way to harden and enforce a state in Kubernetes, then an originalist GitOps tool would be the best fit for them.

Continuous deployment GitOps – originalist and purist

One consideration that the DevOps team had was the possibility of tying in a tool like Argo CD with Ansible or Harness. This integration would allow the teams using Kubernetes to still have the continuous delivery portion automated, but also allow for the self-healing and automated pruning features of Argo CD to enforce the desired end state on the clusters during and after a deployment. But exploring this integration resulted in two main concerns related to inconsistent processes and automated execution led the DevOps team to abandon the idea. They realized that since Argo CD only works on Kubernetes clusters, they would have an inconsistent process with any application that was not Kubernetes-based. Also, Argo CD can only leverage the automated pruning and self-healing features if the automated syncing feature is turned on.

What the team wanted, but couldn't find, was a deployment tool like Argo CD, with the automated pruning and self-healing capabilities, but across all of the different application architecture. Terraform has some capability for this, by leveraging a state file, but the enforcement would only happen on the next execution of Terraform, rather than enforcing the desired configuration immediately. In reality, no other tool has the capability that Kubernetes-based GitOps tools have, such as Argo CD or Flux.

If the core of all GitOps practices is the use of declarative language files that are stored in a source code repository, then any tool that can accomplish or leverage these files becomes a de facto GitOps tool. However, certain tools that advertise themselves as GitOps tools, or are known in the industry as GitOps tools, are typically those tools that are native to Kubernetes, such as Argo CD and Flux. These tools have a reputation of providing low-touch operation, since they are focused solely on enforcing a desired outcome, and what they sacrifice to fulfill the low-touch operation capability is fewer capabilities and options for the end users. The most obvious capability limitation is that the tools only work on one platform. But another limitation is the ability to natively support pre-deployment and post-deployment requirements that a company might need, such as auditing, ticketing, testing, and so on. But, since GitOps tools such as Argo CD and Flux are focused solely on Kubernetes, they fit the most common definition of GitOps in the industry, which is the originalist GitOps practice. Originalist GitOps is a practice that focuses solely on continuous deployment, on a Kubernetes cluster, and leverages only Git repositories as the source of truth.

Expanding a GitOps practice beyond the realm of Kubernetes, but staying focused on continuous deployments, can be considered purist GitOps. Originalist GitOps was coined and championed by Weaveworks a few years ago and has grown in popularity since then. But, because of the rise in other platforms and architecture types, many teams have wanted to extend the originalist GitOps practice to cover more than just Kubernetes. This extension resulted in a less rigid GitOps practice that attempted to get to a purer form and only focus on the core of GitOps, which is declarative files in a source code repository. As a result, any tool that can leverage a set of declarative files to alter the deployment behavior immediately becomes a purist GitOps tool. But the use of a purist GitOps practice means that as long as a script can look through a declarative file and pull values to use in execution, then a script can be a purist GitOps tool.

Although originalist GitOps and purist GitOps practices have some subtle differences, the two important consistencies between them are the use of declarative files and the focus on continuous deployments. But, when a team desires to move beyond continuous deployment and into continuous delivery, that is when verified GitOps comes into view.

Continuous delivery GitOps – verified

One of the hardest parts related to building out the Ansible process was that every action had to be put into code. The DevOps team needed to start with understanding how the Ansible engine would execute the instructions supplied by the declarative files. But as they were researching the underlying mechanisms of Ansible, they realized that there were multiple layers of the declarative files that needed to be built out. At its core, Ansible would take in a set of files that were built out in YAML format, and each key and value pairing would give instructions for Ansible to follow. But as these files needed to be scaled to support different inputs, the team had to then build out a specific type of configuration file, which Ansible calls roles. Those roles would allow for different behavior sets to be passed through the Ansible engine. But with these roles, there were different variations that needed to be considered, which can be accounted for with the use of templating in Jinja, which is the Ansible built-in templating style. After the roles were templated, the files that contained the values to pass to the templates had to be declared in Ansible files so that Ansible could pull the values files. Once the files were pulled, Ansible would then pass the values from the files to fill in the desired templated roles, which would then be passed to the Ansible playbook to be run. To make this abstracted process even more difficult, the DevOps team wanted to allow engineers to supply their own application configuration files to a core set of templates, such as a Helm chart and values file. And added to those layers of declarative files was the use of plugins, which would carry their own YAML file requirements. To finish all of these requirements, Ansible would need a high-level orchestration file that would pull everything together to execute an end-to-end delivery pipeline.

Harness was not much better in regard to what Ansible requires. Every integration in Harness is a configuration file, as is every platform, template script, artifact information, environment definition, deployment process, trigger configuration, and delivery pipeline. What made Harness a bit easier for the DevOps team was that they could build their delivery requirements out in the application interface and the declarative files would be generated as a result. This allowed the DevOps team to build their requirements and then leverage the declarative files for scaling and quicker onboarding. But regardless of Harness, or Ansible, or any other tool, the DevOps team knew that they wanted to leverage the benefits of GitOps across their entire delivery practice. Although they had to go through multiple business requirement changes, platform support scope creep, and a few significant failures, they were able to understand and document their delivery and deployment best practices, which made adopting GitOps much easier.

Any DevOps practice must include continuous integration, continuous deployment, and continuous delivery. As has been discussed thoroughly throughout this book, continuous delivery and continuous deployment are two separate things that must work in tandem to have a truly successful DevOps practice. Adding to the desire for a successful DevOps practice is the desire to automate as much of that practice as possible, if not all of that practice. With a desire to automate comes the inherent requirement of repeatability and reliability. If an automated process is repeatable, meaning that it can be executed repeatedly, but it is not reliable, then the results will be frequent undesired outcomes with a large lack of confidence. And if the automated process is very reliable but cannot be repeated, then the process will have the desired outcome only once.

However, even if an automated process is reliable and repeatable, to be truly effective it must be able to scale. Most non-scalable automated processes that are reliable and repeatable are very narrow in their capabilities and desired outcomes. They might have an ability to leverage different values to change parts of how they execute, but if the process needs to be applied to an entirely different use case, then the process will either need to be rebuilt, or a clone of the process will need to be created and customized accordingly.

Verified GitOps is a GitOps practice that aims to make an automated continuous delivery process reliable, repeatable, and scalable. The way that this can be accomplished is by translating every integration, interaction, action, and outcome, if possible, into declarative code files. For example, if a Helm chart is stored in a Git repository, the Docker image for which is in Docker Hub, and needs to be deployed to three different Kubernetes clusters, creating and updating documentation tickets for each deployment and leveraging an approval gate before deploying to production, then each of those requirements must be defined in code. The integration to the Git repository must be defined in code, as should the file path to the Helm chart. The integration to Docker Hub should also be defined in code, as should the connection requirements to each Kubernetes cluster, with the desired deployment commands. The integration with the ticketing system needs to be defined in code, the ability to create or update a ticket should be defined in code, and so should the ability to wait for an approval.

Defining the different requirements and execution steps in code will result in a reliable process, as the files are what define the behavior of each execution. To make the process repeatable, a high-level object must be defined in code, often called a pipeline. Pipelines can mix and match the different declarative files to achieve a desired outcome, which, when defined in code, becomes a repeatable process that inherits reliability. To make the different pipelines, integrations, and execution steps scalable, every possible value should be templatized to allow for different values to be supplied at any time. Finally, to turn the repeatable, reliable, and scalable process into an automated process, a trigger must be implemented that automatically executes a desired pipeline. Verified GitOps should not be considered only as a practice of translating every continuous delivery requirement into declarative code files that are stored in a Git repository; it also requires that the continuous delivery process is automatically executed. However, before a GitOps practice for either a continuous deployment or continuous delivery process can be implemented, it is important to define best practices for each.

Best practices first, then GitOps

The most important step that the DevOps team had taken over the months of testing different solutions was the process mapping. They had spent many hours interviewing different teams on what their requirements and common activities were, in the hope of being able to automate most, if not all, of it. The interviews covered the engineering teams, the quality assurance and testing teams, the security and compliance teams, and different persons in leadership. As the requirements and individual processes were documented and added together, the DevOps team had to sift through the resulting list and separate out the hard requirements from the nice-to-haves. The hard requirements were any that would prevent a deployment from being successfully rolled out, such as security scanning, dependency deployments, and compliance requirements. The rest of the requirements, such as testing, ticketing, approvals, and documentation, could be handled in a manual way until the DevOps team could add them in. But even though they had the different requirements documented, the delivery and deployment processes were not mapped out in a way that made it clear how one action would lead to the next. For each of the manual steps in the process, the team needed to figure out how to immediately notify the appropriate teams and set a maximum time for the manual step to take before it automatically failed. Every automated step would need some trigger to start it, but since there were multiple manual steps throughout the process, there would need to be an equal number of triggers.

*The other issue that the DevOps team wanted to address was how they could implement a desired process where the current process was lacking. They wanted a continuous process, meaning it would be executed at least once a day, and some of the current steps were not conducive to that desired outcome. Processes such as waiting for the **change approval board**, or **CAB**, would often take many hours, if not days, to get the required approvals. But the steps required to operate without the CAB would not be easy to implement. Automation was a desired outcome for the whole of the engineering organization, not just the DevOps team. But being able to achieve automation seemed like an immovable obstacle.*

In order to get to a fully automated process, the first thing that the team needed to understand was how to implement and enforce best practices for every individual action and requirement. If the best practice was industry-defined, then the team would find easy ways to enforce it. But if there were no industry-defined best practices, then the team would need to find common industry practices and request that someone in leadership would support the enforcement of that common practice. It was paramount that these best practices were defined and enforced first. By defining the many different best practices, the DevOps team found that the entire process became a best practice as a result. After the documenting process was completed, the team could then build out an enforcement of the best practices, to ensure each one was followed. By creating these standards, the DevOps team found that they could easily adopt a GitOps practice, since translating the best practices into declarative code was fairly easy. But if they did not walk through the best practice defining and building process, then any GitOps implementation would have been too difficult, too brittle, and would have lacked enough user adoption to make it successful.

Throughout this book, there has been an emphasis on understanding the underlying requirements of a practice before implementing it. A common reason for implementing any tool or process is often an issue that is currently being faced or one coming in the near future. For example, if there are recent outages for a customer-facing application, a new performance monitoring solution might be purchased to avoid those issues in the future. Another example would be a company trying to add a new compliance certificate, and a new auditing or ticketing tool is implemented to assist with it. Although there are good reasons to implement tools like the ones mentioned, if there is no analysis of the underlying processes or requirements first, then the wrong tool can be implemented, or a tool can be implemented in the wrong way. But the more likely scenario is that what seems like the issue might not be the root cause, but just an effect. To revisit the outage analogy, although a performance monitoring tool is a good tool to have, the issue might be related to a security problem from outside actors, or maybe the issue is actually related to a change in the underlying code or infrastructure. Although monitoring the performance of an application can assist in detecting performance problems, the real cause is outside of the application. Understanding the root cause of the outage would lead to implementing the right tool or process that will mitigate the issue from happening again. To go a bit deeper into the analogy, if the root cause is found, but no analysis is done on similar potential issues, then a tool could be implemented to solve the cause but miss the next issue entirely because of poor implementation.

Best practices do not guarantee perfect outcomes, but rather are intended to show how to avoid issues that stem from improper practices. If a tool is intended to monitor the performance of a server, but the tool is implemented on a database, then best practices are not being followed and there will be false confidence. In the case of GitOps, if the best practices around continuous deployment and continuous delivery are not understood, documented, and implemented, then the GitOps practice will not only fail, but it will also expose all of the issues that arise from improper practices. For example, if a GitOps practice is intended to assist with or even enforce continuous deployment, but the deployment practice is not able to handle daily deployments, then the GitOps practice becomes more of a bottleneck that becomes easily clogged with pending deployments. To achieve the most desirable outcome, it is important to do the right amount of analysis, leverage the right tool or process, and abide by the defined best practices. Only then can a GitOps practice be properly implemented with the desired results.

Summary

You have made it to the end of the book!

This chapter covered a summary of the major topics of the book, such as what continuous practices are; delivery versus deployment; originalist, purist, and verified GitOps; and the importance of best practices.

I hope that you have enjoyed reading this book and that you have gained insight into DevOps, GitOps, and everything in between. Lastly, I hope that this book provided a clearer picture of what is required to achieve repeatability, reliability, and scalability in your continuous delivery and continuous deployment practices through GitOps.

I cannot thank you enough for spending your time reading this book!

Packt.com

Subscribe to our online digital library for full access to over 7,000 books and videos, as well as industry leading tools to help you plan your personal development and advance your career. For more information, please visit our website.

Why subscribe?

- Spend less time learning and more time coding with practical eBooks and Videos from over 4,000 industry professionals

- Improve your learning with Skill Plans built especially for you

- Get a free eBook or video every month

- Fully searchable for easy access to vital information

- Copy and paste, print, and bookmark content

Did you know that Packt offers eBook versions of every book published, with PDF and ePub files available? You can upgrade to the eBook version at packt.com and as a print book customer, you are entitled to a discount on the eBook copy. Get in touch with us at customercare@packtpub.com for more details.

At www.packt.com, you can also read a collection of free technical articles, sign up for a range of free newsletters, and receive exclusive discounts and offers on Packt books and eBooks.

Other Books You May Enjoy

If you enjoyed this book, you may be interested in these other books by Packt:

Learning DevOps

Mikael Krief

ISBN: 978-1-83864-273-0

- Become well versed with DevOps culture and its practices
- Use Terraform and Packer for cloud infrastructure provisioning
- Implement Ansible for infrastructure configuration
- Use basic Git commands and understand the Git flow process
- Build a DevOps pipeline with Jenkins, Azure Pipelines, and GitLab CI
- Containerize your applications with Docker and Kubernetes
- Check application quality with SonarQube and Postman
- Protect DevOps processes and applications using DevSecOps tools

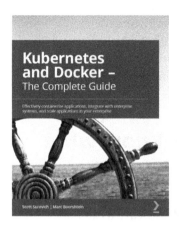

Kubernetes and Docker - An Enterprise Guide

Scott Surovich, Marc Boorshtein

ISBN: 978-1-83921-340-3

- Create a multinode Kubernetes cluster using kind

- Implement Ingress, MetalLB, and ExternalDNS

- Configure a cluster OIDC using impersonation

- Map enterprise authorization to Kubernetes

- Secure clusters using PSPs and OPA

- Enhance auditing using Falco and EFK

- Back up your workload for disaster recovery and cluster migration

- Deploy to a platform using Tekton, GitLab, and ArgoCD

Packt is searching for authors like you

If you're interested in becoming an author for Packt, please visit `authors.packtpub.com` and apply today. We have worked with thousands of developers and tech professionals, just like you, to help them share their insight with the global tech community. You can make a general application, apply for a specific hot topic that we are recruiting an author for, or submit your own idea.

Leave a review - let other readers know what you think

Please share your thoughts on this book with others by leaving a review on the site that you bought it from. If you purchased the book from Amazon, please leave us an honest review on this book's Amazon page. This is vital so that other potential readers can see and use your unbiased opinion to make purchasing decisions, we can understand what our customers think about our products, and our authors can see your feedback on the title that they have worked with Packt to create. It will only take a few minutes of your time, but is valuable to other potential customers, our authors, and Packt. Thank you!

Index

A

B

C

www.ingramcontent.com/pod-product-compliance
Lightning Source LLC
Chambersburg PA
CBHW060521060326
40690CB00017B/3338